SpringerBriefs in Electrical and Computer Engineering

More information about this series at http://www.springer.com/series/10059

Zhiyong Feng • Qixun Zhang • Ping Zhang

Cognitive Wireless Networks

 Springer

Zhiyong Feng
Key Laboratory of Universal Wireless
Communications, Ministry of Education
School of Information and
Communication Engineering
Beijing University of Posts and
Telecommunications
Beijing
China

Ping Zhang
State Key Laboratory of Networking and
Switching
Beijing University of Posts and
Telecommunications
Beijing
China

Qixun Zhang
Key Laboratory of Universal Wireless
Communications, Ministry of Education
School of Information and
Communication Engineering
Beijing University of Posts and
Telecommunications
Beijing
China

ISSN 2191-8112   ISSN 2191-8120 (electronic)
SpringerBriefs in Electrical and Computer Engineering
ISBN 978-3-319-15767-2   ISBN 978-3-319-15768-9 (eBook)
DOI 10.1007/978-3-319-15768-9

Library of Congress Control Number: 2015935656

Springer Cham Heidelberg New York Dordrecht London

Printed on acid-free paper

Springer is a brand of Springer International Publishing Switzerland
Springer is part of Springer Science+Business Media (www.springer.com)

Recommended by Xuemin (Sherman) Shen

Contents

1 Introduction 1

2 Theoretical Study in CWNs 11

3 Novel Architecture Model in CWNs 25

4 Cognitive Information Awareness and Delivery 47

5 Intelligent Resource Management 85

6 TD-LTE Based CWN Testbed 107

7 Standardization Progress 123

8 Conclusion and Future Research Directions 139

Chapter 1
Introduction

With the explosive surge of various applications and high data rate services, different wireless networks and communication technologies are proposed and developed in recent years. Considering the limited radio spectrum resources in various wireless networks, the scarcity of radio spectrum is becoming a bottleneck in face of the exponential surge of service demands. At the same time, spectrum measurement results depict that the average spectrum utilization over a period of time at different locations is quite low, leading to a waste of the valuable spectrum resources. Therefore, how to improve the efficiency of spectrum utilization is the first challenge to be solved for different wireless networks deployment. Besides, the heterogeneity and coexistence of different wireless networks will cause the low radio resource usage and mutual interference among heterogeneous networks due to the lack of cooperative control and information sharing among various networks in practice. Therefore, another big challenge is how to realize the seamless and efficient convergence of different heterogeneous networks to improve the end-to-end performance of users in wireless networks.

To solve these challenges, the heterogeneous wireless networks should be much more intelligent with adaptively reconfigurable parameters and working modes, in order to be aware of the changing wireless network environment. Therefore, the intelligent environment awareness and radio resource utilization technologies should be applied to obtain and analyze the network information learnt from the knowledge representing the dynamics and changing characteristics of radio environment, network traffic and various user demands. Thus, a cognitive wireless network (CWN) is proposed as a novel wireless network enabled by cognitive information awareness, analysis and management technologies, which can improve the efficiency of spectrum utilization and heterogeneous networks convergence.

© The Author(s) 2015
Z. Feng et al., *Cognitive Wireless Networks*, SpringerBriefs in Electrical and Computer Engineering, DOI 10.1007/978-3-319-15768-9_1

1.1 Challenges in Cognitive Wireless Networks

This section introduces three critical challenges faced by wireless communication in detail, which can promote the development of cognitive wireless networks.

1.1.1 Spectrum Scarcity and Spectrum Waste

As one of the most precious non-renewable resources, spectrum resources are licensed and managed by the government, which are facing a big challenge of spectrum scarcity for ubiquitous wireless applications. The feature of spectrum management policy is that a certain part of the spectrum is allocated to dedicated service, and meanwhile, other services are prohibited to utilize this part of spectrum. That is to say, the spectrum is assigned to license holders for a long term basis using a fixed spectrum assignment policy [5]. Figure 1.1 shows the situation of frequency allocation in the United States (U.S.). In Fig. 1.1, the spectrum from 3 KHz to 300 GHz has been allocated completely, which means that there is a small part of the spectrum which can be licensed to new wireless applications [2].

In the U.S., between 2004 and 2005, the Federal Communication Commission (FCC) and Shared Spectrum Company (SSC) had made a survey [3] about the spectrum utilization of 30 MHz~3 GHz in Chicago and New York City. As shown in Fig. 1.2, the survey illustrated that in 2005 long-term spectrum utilization rate was only 5.2% in Chicago and 13.1% in New York. Some spectrum bands were

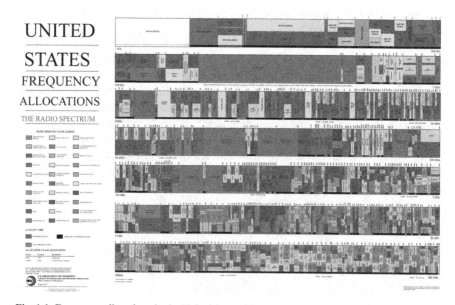

Fig. 1.1 Frequency allocations in the United States [6]

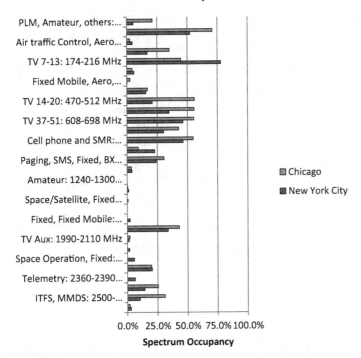

Fig. 1.2 Average spectrum occupancy by band—Chicago vs. New York [3]

overloaded while others were in the state of a low utilization, such as the spectrum band that assigned to radio astronomy.

In Europe, the radio spectrum utilization measurements have been carried out in three different locations, namely, in the suburb of the city of Brno, in the Czech Republic and in the suburb and the city of Paris in France, respectively [7]. The result of the measurement, shown in Fig. 1.3, analyzed the radio spectrum from 400 MHz to 3 GHz.

In China, the measurement results unveiled by [8] are shown in Fig. 1.4, which indicate a low spectrum utilization in Beijing over one month. The results in Beijing are similar to the results released by FCC.

The measurement results above show that some spectrum resources are heavily used by licensed systems in a specific location at a particular time. However, there are many spectrum bands which are only partly occupied or largely unoccupied. Besides, new services and applications need new spectrum resources which is an urgent problem and the bottleneck for the future wireless network. Therefore, how to improve the vacant spectrum utilization and solve the spectrum scarcity problem are key challenges to be addressed. Therefore, new technologies should be utilized in order to detect the vacant spectrum resources efficiently in different locations and

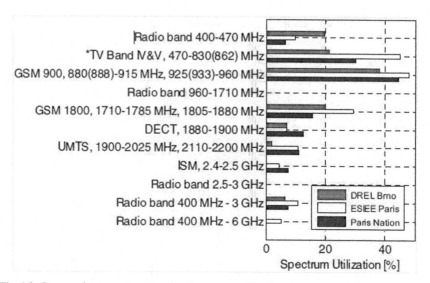

Fig. 1.3 Comparative summary on regional spectrum utilization [7]

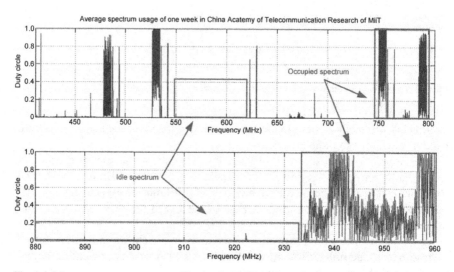

Fig. 1.4 Measurement on spectrum utilization in Beijing [8]

dynamically utilize radio resources. Cognitive wireless network is a candidate solution with intelligent spectrum sensing and dynamic resource management abilities, which can solve the vacant spectrum utilization and spectrum scarcity problems.

1.1.2   Heterogeneous Networks Isolation Problem

Historically, different operators are granted different licensed spectrums and operate different wireless networks with different standards, which leads to the co-existence of multiple radio access technologies (RATs) that may affect each other or sometimes induce excessive interference. In China, different RATs are utilized to provide various services with different quality of services to users, such as GSM, TD-SCDMA, WCDMA, CDMA 2000, WLAN, TD-LTE and so on. Since each network operator plans its own network's coverage independently, coverage holes and excessive coverage problems will exist and are hard to be solved among neighboring cells. Thus, the coverage and interference problems are even more serious in the densely-deployed small cell networks in hot spots and indoor scenarios. Due to this fact, different wireless networks using adjacent spectrum resources will interfere with each other, which seriously harm the overall spectrum usage efficiency of heterogeneous networks and the service experience of end users. Because different heterogeneous networks are lack of control information sharing algorithms and technologies, the cooperation and coordination among heterogeneous networks are difficult to realize and will result in the coverage and interference problems among different networks as big challenges unsolved.

To solve these problems, the cognitive wireless network is proposed to improve the coordination among different heterogeneous networks and promote the convergence of heterogeneous networks. As an efficient control information interworking technology, the cognition information flow technology is also designed to improve the interaction and information exchange among heterogeneous networks, which plays an important role to support the convergence of heterogeneous networks.

1.1.3   Uneven Information Density Distribution and Capacity Demand Surge

Service demands are distributed unevenly at different locations (such as the urban city, suburban area, hotspot, etc.) at different time (fluctuation from day to night), due to the population density distribution pattern and commuters from home to work. Base stations are also deployed with different densities in a large area. Furthermore, deployments of different heterogeneous networks are usually overlapped with each other in hot spot area. And wireless networks with different coverage types, such as macro, pico and femto cells, may be deployed hierarchically to fulfill the capacity demand.

Existing research works mostly focus on the interference mitigation and throughput enhancement techniques from a technical perspective. But how to theoretically define the information density from spatial and temporal perspectives has not been considered yet. Therefore, recent research works on cognitive information metrics, and temporal and geographic distribution entropy theories are proposed and described in this book in order to quantify the unevenly distributed information density in heterogeneous networks.

Furthermore, according to the forecast unveiled by Cisco, mobile data traffic is exploding and doubling each year with the popularization of intelligent terminal and data business growth. Therefore, the future wireless network is preparing for an astounding 1000 times increase of capacity in the next 10 years. How to improve the capacity is a critical issue for wireless researchers and engineers around the world. Obviously, the solution to this formidable challenge is a combination of increasing the efficiency of existing assets, employing more spectrum resources in the form of small cells, as well as adopting radically different ways of acquiring, deploying, operating and managing these resources. However, one of the key challenges is how to identify the unevenly distributed information density in the radio environment. Moreover, another challenge is how to quantify the information density effectively in an efficiently way. Therefore, enabled by cognitive information theory and metrics to solve the unevenly distributed information density, cognitive wireless network will be a good solution to this challenge.

1.2   Overview of Cognitive Wireless Networks

To improve the efficiency of spectrum utilization, cognitive radio (CR) was first proposed by Mitola in 1999 [4], which allows the secondary users (SUs) to opportunistically access the spectrum of the primary users (PUs) without causing any interference. As shown in Fig. 1.5, the spectrum can be dynamically and flexibly shared between PUs and SUs in an opportunistic manner by utilizing the spectrum holes in time and frequency domains, so that the spectrum utilization can be improved.

Recently, driven by the technology innovations and application requirements, wireless network technologies are developed rapidly with high data rate and wide system bandwidth. As different radio access technologies are deployed extensively in metropolises and rural areas, the heterogeneous network coexistence is a big challenge, which leads to interconnection of information, as shown in Fig. 1.6. Therefore, how to realize heterogeneous network convergence and how to improve spectrum utilization are critical issues in future wireless network.

Fig. 1.5 Typical scenario for use of a CR [1]

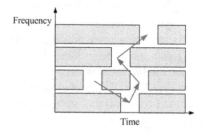

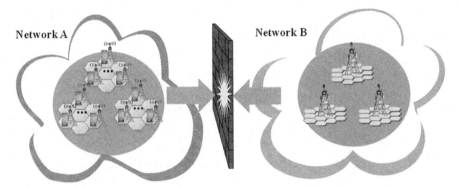

Fig. 1.6 Heterogeneous network

Considering the challenges of network heterogeneity and spectrum inefficient usage, the future wireless networks should have the intelligent network information awareness, flexible spectrum management and dynamic network reconfiguration abilities. This motivates the emergence of Cognitive Wireless Network (CWN). It is defined as a wireless network employing technology to obtain knowledge of its operational and geographical environment, established policies and its internal state; to dynamically and autonomously adjust its operational parameters and protocols according to its obtained knowledge in order to achieve end-to-end network objectives; and to learn from the results obtained.

Therefore, the purpose of this book is to provide a comprehensive introduction to the fundamental theories, disseminate cutting-edge research results, highlight research challenges and open issues, and describe the implementation scheme of the TD-LTE based CWN. The main parts of this book include the theoretical and functional architecture of CWN, cognitive abilities, autonomous decision making techniques and testbed implementation.

1.3   Outline of This Brief

In face of the dynamically changing wireless environment and exponential surge of various service demands, existing research works have not considered about the fundamental theories on how to represent and quantify the dynamic changing characteristics of multi-domains wireless networks with cognitive techniques. In order to enhance the intelligent environment awareness ability in heterogeneous networks, the cognitive information flow theory has been proposed first to identify the relationship between the uncertainty of the network environments and the mutual information awareness removed via the cognitive technology. Furthermore, existing wireless network architecture need to be revolutionarily changed to support the cooperation among different heterogeneous networks enabled by the cognitive information flow and resource flow technologies. Moreover, both the key

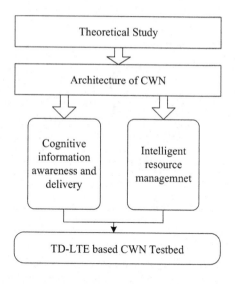

technologies of cognitive information awareness and intelligent resource management are designed to improve the efficiency of spectrum utilization and heterogeneous network convergence. This book consists of five main parts, which addresses a variety of fundamental theory problems and key technologies in CWN. The structure of this book is shown in Fig. 1.7, of which the overall skeleton is from theory to implementation.

Chapter 1 introduces the origin of CWN. Many studies show that while some spectrum bands are extensively used, and most of other bands are largely unoccupied. These potential spectrum holes result in the under-utilization of available bands. Hence, a novel technology to improve the spectrum utilization is proposed. Cognitive radio using dynamic spectrum access technology is considered as an efficient technology to solve this challenge. However, extending to the network perspective, the existing cognitive radio network lacks consideration of the heterogeneous network and the end-to-end performance. So, a novel architecture of CWN is proposed.

Chapter 2 introduces the cognitive information concept to characterize the information sequence in the radio environment awareness results, which applies several novel information theoretic concepts, namely, cognitive information, geographic entropy and temporal entropy to reveal the features of cognitive information.

Chapter 3 focuses on the architecture of CWN. First, some classical cognitive network architecture will be discussed, such as cognitive cycle of Mitola and basic cognitive cycle of Haykin, and then functional architecture proposed by IEEE 1900.4 work group will be illustrated. Second, the theoretical architecture and functional architecture are proposed, through analyzing the merits and drawbacks of previous architectures. Furthermore, in order to improve the network capacity of the proposed architecture, a heterogeneous network architecture based on the hierarchical deployment is proposed, which can effectively improve the resource utilization, reducing the network overhead. Finally, some key concepts and metrics will be introduced, which are the fundamental of the novel architecture.

Cognitive information awareness is the first step of a CWN in gathering the necessary network information such as available spectrum and network operation parameters. In Chap. 4, the cognitive ability is proposed which can be mainly categorized as spectrum sensing, cognitive pilot channel (CPC) and cognition database according to different categories of information and collection methods.

Chapter 5 focuses on intelligent resource management. CWN imposes challenges due to the fluctuating nature of the available spectrum, as well as the diverse QoS requirements of various applications. Spectrum management functions can address these challenges for the realization of this new network paradigm. In this part, there are three types of management: Dynamic Spectrum Management (DSM), Joint Radio Resource Management (JRRM), and Transmit Power Control (TPC).

Chapter 6 introduces the CWN Testbed. As a verification of theoretical research and a feasible prototype, a CWN testbed is designed and developed in Time Division Long Term Evolution (TD-LTE) cellular system at Beijing University of Posts and Telecommunications (BUPT), which is sponsored by the National Basic Research Development Program of China, the National Natural Science Foundation of China, the National Science and Technology Major Project and the Program for New Century Excellent Talents in University.

Chapter 7 introduces the standard progress of CR technology, and three main standardization organizations, International Telecommunications Union-Radio Communications Sector (ITU-R), Institute of Electrical and Electronic Engineers (IEEE) and European Telecommunications Standards Institute (ESTI) make contributions to the regularity of CR application.

The conclusion and future research directions are given in Chap. 8.

References

1. Akyildiz IF, Lee WY, Vuran MC (2008) A survey on spectrum management in cognitive radio networks. IEEE Commun Mag 46(4):40–49
2. Ekram H, Dusit N, Zhu H (2009) Dynamic spectrum access and management in cognitive radio network. Cambridge University Press, New York
3. Mark AM, Peter AT, Dan M (2006) Chicago spectrum occupancy measurements & analysis and a long-term studies proposal. Paper presented at the Proceedings of the first international workshop on technology and policy for accessing spectrum, New York, 2006
4. Mitola J (1999) Cognitive radio: making software radios more personal. IEEE Pers Commun 6(4):13–18
5. Simon H (2005) Cognitive radio: brain-empowered wireless communications. IEEE J Sel Areas Commun 23(2):201–220
6. Spectrum Dashboard (2010) REBOOT FCC GOV, Washington. http://reboot.fcc.gov/spectrumdashboard/searchSpectrum.seam. Accessed 17 Nov 2010
7. Vaclav V, Roman M, Genevieve B et al (2010) Survey on spectrum utilization in Europe: measurement, analyses and observation. CROWNCOM, Cannes, 9–11 June 2010
8. Zhang P, Liu Y, Feng Z Y et al (2012) Intelligent and efficient development of wireless networks: a review of cognitive radio networks. Chin Sci Bull 57(28–29):3662–3676

Chapter 2
Theoretical Study in CWNs

In this chapter, the latest information theory based theoretical results in CWNs are introduced in detail. Compared to existing works on physical layer spectrum sensing and signal processing techniques, the cognitive information concept has been proposed to characterize the information sequence from radio environment awareness results. And the uncertainty of radio environment is depicted by novel definitions of geographic entropy and temporal entropy. Furthermore, the effectiveness of cognitive techniques which can eliminate the uncertainty of radio environment are verified by the proposed cognitive information metrics.

2.1 Theoretical Challenges in CWNs

Driven by technology innovations and various service requirements, new wireless network technologies with high data rate and wide spectrum band are developing rapidly. However, radio spectrum resources are facing spectrum scarcity and spectrum underutilization problems. Therefore, cognitive radio (CR) [1] has been proposed as one of the most promising technologies for the efficient spectrum utilization in recent years. With the flexible and comprehensive usage of available spectrum resources in [2] and [3], CR enables the optimal and efficient radio resource utilization. As Haykin in [1] pointed out, CR is an intelligent wireless communication system aware of its surrounding environment. Exploiting the information of radio environment with the environment awareness technologies, CR is able to learn and reconfigure itself to adapt to its environment. In this chapter, the cognitive information of radio environment by using information theoretic techniques is studied and the compression of cognitive information is proposed in CWNs.

Since the emergence of CR, the exploration and exploitation of the spectrum opportunities are extensively studied in the literatures [4]. Besides, advanced techniques such as machine learning is also applied in the network with cognitive

© The Author(s) 2015
Z. Feng et al., *Cognitive Wireless Networks,* SpringerBriefs in Electrical
and Computer Engineering, DOI 10.1007/978-3-319-15768-9_2

radio equipments to obtain the states information of primary users (PUs), which is essential for the dynamic spectrum access (DSA). However, the existing works mainly focus on the signal processing in spectrum sensing, while the characteristics and compression of cognitive information are ignored in the research of radio environment awareness technologies such as spectrum sensing. Therefore, the cognitive information concept has been proposed in this chapter to characterize the information sequence in the radio environment awareness results by using some novel information theoretic concepts, namely, cognitive information, geographic entropy and temporal entropy to reveal the features of cognitive information.

2.2   Information Theory Applied in Cognitive Wireless Networks

The information theories have been applied in spectrum sensing by researchers recently. The most representative literature is shown in [5] and [6], where the authors use the information theoretic criterion to detect the existence of PUs. Wang et al. in [5] applied the Akaike Information Criterion (AIC) and the minimum description length (MDL) criterion in the blind spectrum sensing. Liu et al. in [6] discovered the fact that the entropy of noise is smaller than the entropy of noise plus signal, thus the entropy based criterion can be used in spectrum sensing. Notice that the information theoretic techniques in previous literatures are mainly in the area of signal processing. The study of sensing-throughput tradeoff [7] can be regarded as the study of cognitive information. However, all these literatures do not explicitly address the behavior and characteristics of cognitive information and the information compression issues. As far as we know, the study of cognitive information is still very limited.

2.3   Geographic and Temporal Distribution Entropy Theory

Facing the challenges of time-variant wireless channel, complex network radio resource management and a variety of user behaviors in wireless network environments, wireless networks need to implement the multi-domain environment cognition technologies, such as the spectrum sensing to discover vacant radio resources in the wireless domain, database based environment cognition technology to discover the vacant resources in the network domain and user domain. In order to provide a universal measurement for the cognitive information, information theoretic method is designed and proposed in this chapter. And cognitive information is defined as the uncertainty of the network environment that cognition technology can remove via the environment awareness techniques, such as spectrum sensing. Therefore, cognitive information is defined as the mutual information.

In cognitive wireless networks, the radio environment, also named by the wireless domain, is monitored by spectrum sensing equipments. If the radio environment is complex, the intensity of information must be high enough to eliminate the uncertainty of the radio environment. In this section, the uncertainty of radio environment is defined and studied. Besides, the notions of geographic entropy and temporal entropy are proposed to measure the uncertainty of radio environment in space and time dimensions respectively. Then, the relation between the intensity of cognitive information and the uncertainty of the radio environment is studied thereafter. In order to quantify the complexity of wireless networks coverage in different locations, the geographic entropy based Radio Environment Information (REI) and Radio Parameter Error (RPE) representation approaches are proposed respectively.

2.3.1   REI Representation

Cognitive radio allows the optimization of radio resource utilization by exploiting vacant spectrum resources, such as the spectrum holes, which exist in the space and time dimensions. Considering the complex radio environment information caused by the difference of wireless network distributions, the radio environment information (REI) is defined as a parameter representing the available radio resources at a specific location which can be utilized by CWNs. And the characteristics of the REI are investigated in the spatial dimension. As illustrated in Fig. 2.1, there are three primary base stations (BSs), when a cognitive radio is in the coverage of primary BSs, the resources of primary BSs are not available for this cognitive radio. However, cognitive radios can exploit the spatial spectrum holes when it is outside of the coverage of primary BSs.

Fig. 2.1 Scenario of multiple primary networks

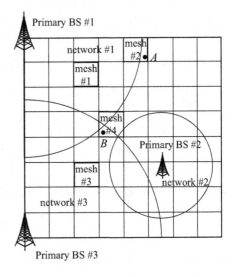

To represent the REI, $R(k,x,y)$ is defined as the binary representation of network k at location (x,y) in Eq. (2.1). The radio parameter at a specific location is denoted by the sum of the binary representations for all networks in Eq. (2.2) based on the assumptions in [8, 9], where T is the number of networks. The entire region is divided into uniform small squares, which is denoted as "mesh". Moreover, different meshes represent various coverage conditions of multiple radio access networks. The radio parameter of mesh i is denoted in Eq. (2.3), where p_{ij} is the fraction of the area in mesh i with radio parameter j, $\sum_{j=1}^{N} p_{ij} = 1, N = 2^T$ is the number of radio parameters.

In Fig. 2.1, there are $T = 3$ primary networks and the number of radio parameters is $N = 2^T = 8$. Therefore, the radio parameter in mesh 2 is 1 according to Eq. (2.3).

$$R(k,x,y) = \begin{cases} 1 \text{ if network } k \text{ is detected at } (x,y) \\ \qquad 0 \text{ otherwise} \end{cases} \tag{2.1}$$

$$I(x,y) = \sum_{k=1}^{T} R(k,x,y) \times 2^{k-1} \tag{2.2}$$

$$P = \arg\max_{j} p_{ij} \tag{2.3}$$

2.3.2   Geographic Entropy

Based on the assumption that each mesh represents the most commonly available network information in its specific location, the radio parameter error (RPE) is defined in Eq. (2.4) and the RPE of the entire region is defined in Eq. (2.5), where M is the number of mesh and α_i is the proportion area of mesh i comparing to the entire region which satisfies $\sum_{i=1}^{M} \alpha_i = 1$. The Geographic Entropy (GE) of a mesh is defined to quantify the uncertainty of the radio environment information, which is caused by the difference of wireless network distributions. In Fig. 2.1, the REI in mesh 1 is much more certain than that in mesh 4 in terms of the distribution of different networks. The REI in mesh 4 is more composite and complex, causing a much bigger uncertainty. Similar to the Shannon entropy [10], the geographic entropy of mesh i is denoted by H_i in Eq. (2.6). And the geographic entropy of the entire region is defined by H in Eq. (2.7). For a uniform size mesh division approach, $\alpha_i = \dfrac{1}{M}$ and H are shown in Eq. (2.8).

$$p_{e,i} = 1 - \max_{j} p_{ij} \tag{2.4}$$

$$p_e = \sum_{i=1}^{M} \alpha_i p_{e,i} \tag{2.5}$$

$$H_i = -\sum_{j=1}^{N} p_{ij} \log p_{ij} \tag{2.6}$$

$$H = \sum_{i=1}^{M} \alpha_i H_i = -\sum_{i=1}^{M} \alpha_i \sum_{n=1}^{N} p_{ij} \log p_{ij} \tag{2.7}$$

$$H = \frac{1}{M} \sum_{i=1}^{M} H_i \tag{2.8}$$

Therefore, by using the geographic entropy concept, the uncertainty of the REI and the corresponding RPE values are quantified, which can be applied to analyze the relation between the number of mesh M and RPE. It can be applied in the design of new algorithms for the efficient environment information awareness.

Moreover, the properties of geographic entropy and RPE are investigated, considering the relation between M and the RPE based on the proposed geographic entropy concept.

Theorem 1 The GE of the entire region is $O(\frac{1}{\sqrt{M}}) \to 0$, where M is the number of meshes.

Proof The meshes with impure radio environment are distributed along the boundaries of primary networks. Denote the length of all these boundaries as ξ, the length of a mesh edge as ε, and the length of the entire region's edge as L. Then, $M = (\frac{L}{\varepsilon})^2$ and the number of mesh with impure radio environment is upper bounded by K in Eq. (2.9).

$$K \leq \frac{2\xi\sqrt{2}\varepsilon}{\varepsilon^2} = \frac{2\sqrt{2}\xi}{\varepsilon} \tag{2.9}$$

As shown by the solid curve in Fig. 2.2, the result is obtained by considering the corresponding packing problem along the boundaries of detection region of TV primary networks. Moving each point on this curve in two normal directions with a distance of $\sqrt{2}\varepsilon$, two dash curves are achieved hereafter. The area between these

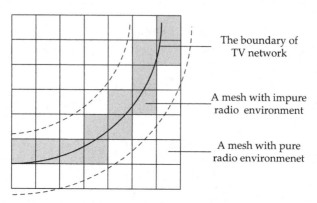

Fig. 2.2 The boundary of detection region determines an upper bound on K

two dash curves is $2\xi\sqrt{2}\varepsilon$. All the meshes with a pure radio environment are located between these two dash curves, so the upper bound on K is $2\xi\sqrt{2}\varepsilon$ divided by the area of a mesh.

$$H \leq \frac{1}{M}K\log N \leq \frac{1}{M}\frac{2\sqrt{2}\xi}{\varepsilon}\log N = \frac{1}{\sqrt{M}}\frac{2\sqrt{2}\xi\log N}{\sqrt{S}} \qquad (2.10)$$

Since $H_i \leq \log N$, thus an upper bound of the geographic entropy is indicated in Eq. (2.10), where S is the area of the entire region. Thus $O\left(\frac{1}{\sqrt{M}}\right) \to 0$ is an upper bound of H.

2.3.3 Temporal Entropy

Considering the proposed geographic entropy to quantify the uncertainty of various wireless networks distribution, the intensity of cognitive information distribution in the time dimension is defined by the cognitive information flow (CIF), which is a dynamic and ordered information sequence of the radio environment sensing results in CWNs. CIF contains the information of spectrum occupancies, namely, the states of spectrum holes. Furthermore, the information in CIF will flow to cognitive radio equipments or the database that stores the spectrum occupancy information.

The state of radio environment depicts the state of PUs (busy or idle). And the state of PUs is defined as the binary representation of spectrum occupancy at time t.

$$I(t) = \begin{cases} 1 & \text{If spectrum band is occupied} \\ 0 & \text{Otherwise} \end{cases} \qquad (2.11)$$

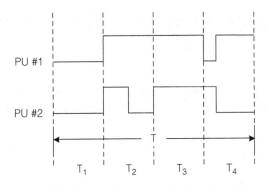

Fig. 2.3 Temporal entropy of two primary users

For the PUs operating on the single spectrum band, $I(t)$ is a "0–1" random variable (r.v.). In Fig. 2.3, the spectrum occupancy of two primary users is observed during a time interval T, which is divided into N time windows as $T_1, T_2, \ldots, T_N$ ($N=4$). If $T_i = T_j, \forall i \neq j$ and the SU detects the spectrum at the beginning of time window $T_i, \forall i$, then T/N is the spectrum sensing period. The state of PU during time window T_i is denoted as X_i (1 or 0), which is a random variable (r.v.) and is defined in Eq. (2.12), where p_{ij} is the fraction of time in time window T_i when the state of PU is $j \in \{0,1\}$.

$$X_i = \arg\max_j p_{ij} \tag{2.12}$$

To capture the dynamic state of PUs, we define the temporal entropy to characterize the uncertainty of the states of spectrum occupancy. The temporal entropy of the entire interval T is defined as follows.

$$H = \frac{1}{N} H(X_1, X_2, \cdots, X_N) \tag{2.13}$$

This is the form of entropy rate, i.e., the average entropy of each symbol. Since different values of X_i are independent, the temporal entropy can be simplified as follows.

$$H = \frac{1}{N} \sum_{i=1}^{N} H(X_i) \stackrel{(a)}{=} \frac{1}{N} \sum_{i=1}^{N} H_i \tag{2.14}$$

Specifically, H_i is the entropy of spectrum occupancies in time window T_i, defined as

$$H_i = -\sum_{j=1}^{M} p_{ij} \log p_{ij} \tag{2.15}$$

where M is the number of states. For spectrum sensing in single spectrum band, $M=2$. p_{ij} is the fraction of time with state j in time window T_i, which satisfies

$$\sum_{j=1}^{M} p_{ij} = 1, \forall i \cdot$$

The error probability of time window T_i is defined in Eq. (2.16), where X_i is defined in Eq. (2.12).

$$p_{e,i} = 1 - \max_{j} p_{ij} = 1 - p_{iX_i} \tag{2.16}$$

Equation (2.16) reveals that even if the spectrum sensing is perfect (no miss detection or false alarm), SU may still make mistakes.

As illustrated in Fig. 2.3, for PU #1, at the beginning of T_4, SU declares that PU #1 is idle, so SU will transmit signals during T_4, and it will interfere PU #1 during most of the remaining time, leading to another case of "miss detection". Similarly, for PU #2, at the beginning of T_4, SU declares that PU #2 is busy, so SU does not transmit signals during the remaining time, while PU #2 is idle for most of the time, resulting in another case of "false alarm". Thus, even though SUs detect the spectrum perfectly, other forms of miss detection and false alarm are inevitable. The error probability is used to measure this kind of mistakes. The error probability of the entire time interval is denoted in Eq. (2.17).

$$p_e = \sum_{i=1}^{N} \frac{T_i}{T} p_{e,i} \overset{(b)}{=} \frac{1}{N} \sum_{i=1}^{N} p_{e,i} \tag{2.17}$$

Similarly, (b) holds when the time windows are equal. As to the mathematical properties of temporal entropy, the theorem is achieved below.

Theorem 2 The temporal entropy is $H = O\left(\dfrac{1}{N}\right)$, where N is the number of time windows and $\lim_{N \to \infty} H = 0$.

2.4   Cognitive Information Metrics

The cognitive information is defined as a metric to evaluate the uncertainty of both internal and external environments. Such uncertainty can be eliminated by other systems or nodes using cognitive awareness techniques. In CWNs, $X(i)$ is used to represent the internal and external environment states of system A at the time instance i. The cognitive awareness result of system A's state by system B using cognitive awareness techniques is denoted by $Y(i)$, then the output sequence can be expressed by $Y(1), Y(2), \cdots, Y(N)$, where N is the number of time instances. The averaged mutual information is defined in Eq. (2.18). It is defined generally as the uncertainty of system A's states which is removed by system B in N times using the cognitive information awareness techniques. However, as the correlation between the states of two time instances is unknown, it is difficult to find a closed-form solution for Eq. (2.18). Therefore, the simplified form of Eq. (2.18) with $N=1$ is achieved and the cognitive information is defined below.

$$\frac{1}{N} I(X(1), X(2), \cdots, X(N); Y(1), Y(2), \cdots, Y(N))$$
$$= \frac{1}{N} H(X(1), X(2), \cdots, X(N)) - \frac{1}{N} H\left(X(1), X(2), \cdots, X(N) \mid Y(1), Y(2), \cdots, Y(N)\right)$$
$$(2.18)$$

Definition 1 Cognitive information is defined as the mutual information,

$$I(X;Y) = H(X) - H(X \mid Y) \qquad (2.19)$$

X is the state of system A and Y is the inference result of X by system B using cognitive awareness techniques. Both X and Y are random variables. $H(X)$ is the entropy of system A's states and $H(X \mid Y)$ is the conditional entropy.

Remark 1 The entropy $H(X)$ is useful in the source coding of the system A's states, which is essential to reduce the cognitive information exchanging among different systems or nodes using cognitive information awareness techniques.

Remark 2 The cognitive information $I(X;Y)$ is defined as the information that one system "transmits" to another system or node, namely, the uncertainty of the states of system A that can be removed by system B using cognitive awareness techniques.

Considering the imperfect cognitive awareness in CWNs $\left(H(X \mid Y) \neq 0\right)$, the error probability exists and can be defined by Eq. (2.20).

$$p_e = \Pr\{X \neq Y\}, \qquad (2.20)$$

Generally, the states of one system or node are complex and changing dynamically, including the radio, network, user and policy multi-domains. For simplicity, it is assumed that the node usually has two states, the idle state and the busy state. For example, the spectrum occupancy state is taken into account and the expression of the error probability is depicted in Eq. (2.21).

$$p_e = p_0 p_f + p_1 p_m, \qquad (2.21)$$

p_0 is the probability that the node is in the idle state "0" and p_1 is the probability that the node is in the busy state "1", p_f is the probability of false alarm and p_m is the probability of miss detection. Thus, the cognitive information is depicted in Eq. (2.22).

$$I(X;Y) = H(X) - H(X \mid Y) = H(Y) - H(Y \mid X)$$
$$= -\sum_{i \in \{0,1\}} q_i \log q_i + \sum_{x \in \{0,1\}} \sum_{y \in \{0,1\}} p(x)p(y \mid x) \log p(y \mid x), \qquad (2.22)$$

And the expressions of q_0, q_1 denote the probabilities of the idle state "0" and busy state "1" of the detection results, as $q_0 = p_0(1 - p_f) + p_1 p_m$ and $q_1 = p_0 p_f + p_1(1 - p_m)$.

In this section, the mathematical features of cognitive information are described in detail by using the cognitive information as a metric for energy detection, cooperative spectrum sensing and network performance evaluation.

2.4.1 Cognitive Information Metrics in Energy Detection

When the system A's signal $x(t)$ is transmitted through wireless channel with channel gain $h(t)$, the received signal from another system or node B using cognitive awareness techniques is $y(t)$. It follows a binary hypothesis: H_0 (when signal $x(t)$ is absent) and H_0 (when signal $x(t)$ exists).

$$y(t) = \begin{cases} w(t) & : H_0 \\ x(t)h(t) + w(t) & : H_1 \end{cases}, \qquad (2.23)$$

And $w(t)$ is the additive white Gaussian noise (AWGN), which is assumed to be a circularly symmetric complex Gaussian random variable with the mean zero and one-sided power spectral density σ_w^2, namely, $w(t) \sim \mathcal{N}(0, \sigma_w^2)$.

The decision statistic is $Y = \dfrac{1}{N} \sum_{i=1}^{N} |y(i)|^2$. With the assumption that $x(n)$ is complex Phase-shift Keying (PSK) modulated and $w(n)$ is circularly symmetric complex Gaussian (CSCG) noise, the probabilities of miss detection $p_m = \Pr\{H_0 \mid H_1\}$ and false alarm $p_f = \Pr\{H_1 \mid H_0\}$ are shown in Eqs. (2.24) and (2.25) in [11].

$$p_m(\varepsilon, N) = Q\left(\left(\gamma + 1 - \frac{\varepsilon}{\sigma_w^2} \right) \sqrt{\frac{N}{2\gamma + 1}} \right), \qquad (2.24)$$

$$p_f(\varepsilon, N) = Q\left(\left(\frac{\varepsilon}{\sigma_w^2} - 1 \right) \sqrt{N} \right), \qquad (2.25)$$

N is the number of samples, γ is the signal to noise ratio (SNR) at the signal receiver node B, σ_w^2 is the power spectral density of AWGN and ε is the decision threshold. $Q(x) = \frac{1}{\sqrt{2\pi}} \int_x^\infty \exp\left(-\frac{t^2}{2}\right) dt$ is the tail probability of the standard normal distribution.

Therefore, the theorem on the cognitive information and error probability is achieved and described below.

Lemma 1 The error probability is $\Theta\left(e^{-\kappa N}\right)$, where N is the number of samples and κ is a constant positive number.

Proof Use the expressions of the probabilities of miss detection and false alarm in Eqs. (2.24) and (2.25). For $x \geq 0$, an upper bound of the Q-function is depicted in Eq. (2.26).

$$Q(x) \leq \frac{1}{2}\exp(-\frac{x^2}{2}), \tag{2.26}$$

According to Eqs. (2.24), (2.25), and (2.26), the results are shown in Eqs. (2.27) and (2.28).

$$p_f(\varepsilon, N) \leq \frac{1}{2}\exp(-\frac{1}{2}(\frac{\varepsilon}{\sigma_w^2} - 1)^2 N), \tag{2.27}$$

$$p_m(\varepsilon, N) \leq \frac{1}{2}\exp(-\frac{1}{2(2\gamma+1)}(\gamma+1-\frac{\varepsilon}{\sigma_w^2})^2 N), \tag{2.28}$$

Use the definition of error probability Eq. (2.21), the results are shown in Eqs. (2.29) and (2.30).

$$P_e \leq \frac{p_0}{2}\exp(-\frac{(\frac{\varepsilon}{\sigma_w^2} - 1)^2}{2} N) + \frac{p_1}{2}\exp(-\frac{(\gamma+1-\frac{\varepsilon}{\sigma_w^2})^2}{2(2\gamma+1)} N), \tag{2.29}$$

$$\kappa = \min\left\{\frac{(\frac{\varepsilon}{\sigma_w^2} - 1)^2}{2}, \frac{(\gamma+1-\frac{\varepsilon}{\sigma_w^2})^2}{2(2\gamma+1)}\right\}, \tag{2.30}$$

Then $\Theta\left(e^{-\kappa N}\right)$ is denoted in Lemma 1.

Theorem 3 The conditional entropy $H(X\,|\,Y)$ is $O\left(Ne^{-\kappa N}\right)$, where N is the number of samples and κ is a constant positive number.

Proof According to Fano's inequality, the result is denoted below.

$$H(X\,|\,Y) \leq H(p_e) + p_e|M-1|, \tag{2.31}$$

By using Lemma 1, $H(p_e) = \Theta\left(Ne^{-\kappa N}\right) + \Theta\left(e^{-\kappa N}\right), p_e = \Theta\left(e^{-\kappa N}\right)$.   Thus $H(X\,|\,Y) = O\left(Ne^{-\kappa N}\right)$.

However, in the practical application scenario, using cognitive information awareness technique, such as the spectrum sensing, the number of samplings N can not be infinite. With a finite N, the detection threshold can be tuned to optimize the performance of energy detection.

2.4.2   Cognitive Information Metrics in Cooperative Awareness

Suppose that there are M nodes to detect a spectrum band occupied by system A and the detection results are $Y_1, Y_2, \ldots, Y_M$, then the uncertainty of system A that is removed by M nodes is shown in Eq. (2.32).

$$
\begin{aligned}
I(X; Y_1, Y_2, &\cdots, Y_M) \\
&= \sum_{i=1}^{M} I(Y_i; X\,|\,Y_{i-1}, Y_{i-2}, \cdots, Y_1) \\
&= I(Y_1; X) + I(Y_2; X\,|\,Y_1) + I(Y_3; X\,|\,Y_2, Y_1) + \cdots \\
&\geq I(Y_1; X)
\end{aligned}
\tag{2.32}
$$

Lemma 2 can be derived from Eq. (2.32).

Lemma 2 The cognitive information achieved by cooperative cognitive awareness technique, such as the cooperative spectrum sensing, is bigger than that of local cognitive awareness technique with a single node.

Definition 2 Cooperative gain in the cooperative cognitive information awareness is defined as the margin between two mutual information.

$$
G = I(X; Y_1, Y_2, \cdots, Y_M) - I(X; Y_1),
\tag{2.33}
$$

As to the relation between cooperative gain G and the number of cooperative nodes M, Lemma 3 is achieved below.

Lemma 3 Cooperative gain is an increasing function of M.

Proof This lemma is proved by Eq. (2.32).

Lemma 3 denotes that the number of cooperative nodes in cooperative cognitive information awareness is in direct proportion to the cognitive information. However, in the K-out of-M criterion of cooperative information awareness, such as the cooperative spectrum sensing, the sensing performance is not an increasing function of K.

2.5 Concluding Remarks

Considering the changing wireless environment and various user demands, the wireless network should be able to be intelligently aware of the environment, dynamically utilize and efficiently management radio resources, which are crucial for enhancing intelligent abilities of the network. In this chapter, the cognitive information is proposed as a universal method to measure the environment cognition results in wireless networks. Besides, both the geographic entropy and temporal entropy are proposed to reveal the uncertainty of the changing radio environment. Meanwhile, the relation between the uncertainty of radio environment and the intensity of cognitive information flow is also studied. Thus, in intelligent wireless networks which needs environment cognition, the proposed cognitive information can not only quantify the cognition capability of wireless networks, but also provide fundamental algorithms to optimize the efficient allocation of cognition resources such as the deployment density of sensors and the period of spectrum sensing in cognitive wireless networks.

References

1. Simon H (2005) Cognitive radio: brain-empowered wireless communications. IEEE J Sel Areas Commun 23(2): 201–220
2. Ekram H, Dusit N, Zhu H (2009) Dynamic spectrum access and management in cognitive radio network. Cambridge University Press, New York
3. Spectrum Dashboard (2010) REBOOT FCC GOV, Washington DC. http://reboot.fcc.gov/spectrumdashboard/searchSpectrum.seam. Accessed 17 Nov 2010
4. Mark AM, Peter AT, Dan M (2006) Chicago spectrum occupancy measurements & analysis and a long-term studies proposal. Paper presented at the Proceedings of the first international workshop on Technology and policy for accessing spectrum, New York, 2006
5. Vaclav V, Roman M, Genevieve B et al (2010) Survey on spectrum utilization in europe: measurement, analyses and observation. CROWNCOM, Cannes, 9–11 June 2010
6. Zhang P, Liu Y, Feng ZY et al (2012) Intelligent and efficient development of wireless networks: A review of cognitive radio networks. Chin Sci Bull 57(28–29):3662–3676
7. Mitola J (1999) Cognitive radio: making software radios more personal. IEEE Pers Commun 6(4):13–18
8. Thomas R, Friend D, DaSilva L et al (2006) Cognitive networks: adaptation and learning to achieve end-to-end performance objectives. IEEE Commun Mag 44(12):51–57
9. Denkovski D, Pavlovska V, Atanasovski V et al (2010) Novel policy reasoning architecture for cognitive radio environments. IEEE GLOBECOM, Miami, 6–10 Dec 2010
10. Mitola J (1993) Software radios: survey, critical evaluation and future directions. IEEE Aero El Sys Mag 8(4):25–36
11. Buljore S, Harada H, Filin S et al (2009) Architecture and enablers for optimized radio resource usage in heterogeneous wireless access networks: the IEEE 1900.4 working group. IEEE Commun Mag 47(1):122–129

Chapter 3
Novel Architecture Model in CWNs

In this chapter, innovative architectures of CWN are introduced in detail. After a brief survey of the research works of existing architectures, a novel architecture of CWN is proposed, which features the cognitive information flow and resource flow. Then, the functional architecture of the proposed novel CWN is also introduced. Finally, the deployment architecture is briefly illustrated from an implementation aspect in terms of the unevenly information density distribution.

3.1 Theoretical Architecture Model

In the past decade, many studies have focused on the network architecture of cognitive radio. Haykin proposed a simple DSA architecture, in reference [9], which emphasized the need for cognitive radio to be aware of its radio environment and avoid interference. In [11], Thomas proposed a cognitive framework for cognitive radio to achieve the end-to-end goals, where a cognitive process layer was introduced to perform the cognitive functionality. In [3], Denkovski presented a policy reasoning architecture for cognitive radio where a variety of policy types for different purposes was allowed for high network flexibility and adaptability.

The research on cognitive radio is mainly derived from the concept of a cognition cycle that is the earliest theoretical architecture model, shown in Fig. 3.1, in which the wireless network processes are mapped according to cognition theory started by Mitola[6]. Reasoning components are incorporated in the architecture to make plans and decisions more intelligent in wireless applications. As shown in Fig. 3.2, the proposed architecture by Haykin in [9] emphasizes on the awareness of the radio environments to operate in primary spectrum bands. The above two architectures are the most popular examples. Besides, the cognition cycle proposed by Virginia Tech emphasizes the functions of reasoning and learning [1].

However, the architectures in the above work only focus on some specific applications of cognitive radio, such as the performance of the wireless transceiver,

© The Author(s) 2015
Z. Feng et al., *Cognitive Wireless Networks,* SpringerBriefs in Electrical and Computer Engineering, DOI 10.1007/978-3-319-15768-9_3

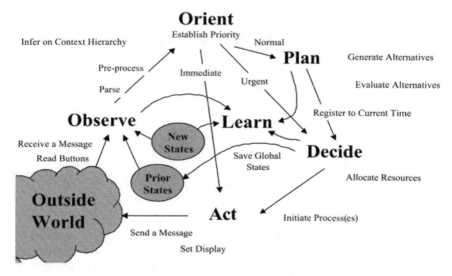

Fig. 3.1 Cognition cycle proposed by Mitola [7]

Fig. 3.2 Cognition cycle
proposed by Haykin [9]

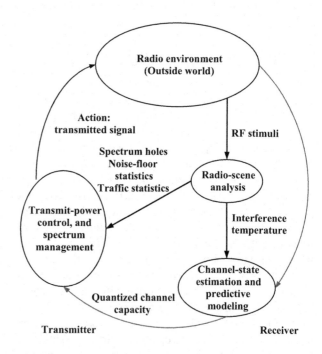

which is not enough to guarantee the end-to-end performance of the CWN from the
network perspective. The main drawbacks are:

1. The work objective of cognition cycle is vague
2. The objective of each sub-cycle is obscure (for example, the function of Observe
   is not defined)

3. The flow direction of cognitive information is indeterminate
4. The interactive relationship of each part is not clear (for example, the Orient should guide the Observe)
5. The conditions for cycle to initiate the termination is uncertain

Therefore, a novel functional architecture for CWN is proposed, which will be introduced in Sect. 3.2. The proposed architecture provides a high level abstraction of CWN, where the end-to-end goals and three key capabilities of CWN: cognition capability, self-organization capability and reconfiguration capability are taken into account. Considering the existing structure models and end-to-end performance goals, a theoretical model of the cognitive wireless network architecture is designed in Fig. 3.3.

The theoretical model of CWN architecture includes several modules:

1. *End-to-end Goal*: This module is the driving goal of the theoretical model of the entire CWN architecture. Cognitive wireless network will collect relevant cognitive information based on the end-to-end goal. Then it will establish the mapping of the environment, form a plan, optimize decision-making, and then perform the reconstruction based on cognitive information. This process will continue until the realization of the end-to-end goal. Once achieving the end-to-end goal, the network will be performed in accordance with the existing planning and decision

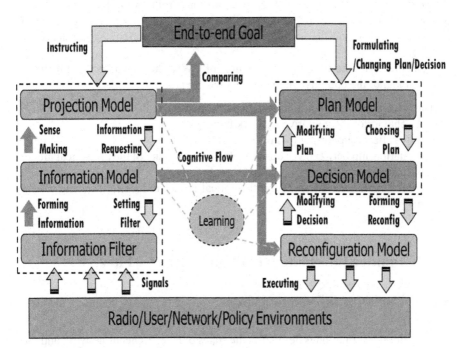

Fig. 3.3 The theoretical model of CWN architecture

making until the multi-domain environment changes lead to a reduction in end-to-end performance, resulting in the need to remake the plan and decision.

2. *Information Filter*: It is mainly used to set filters to obtain the cognitive information from the multi-domain environment.
3. *Information Model*: It is used to establish the cognitive information model for the precise mapping of the multi-domain environment.
4. *Projection Model*: It is used to extract the cognitive information from the Information Model to establish a multi-domain environment mapping. Furthermore, it can achieve the end-to-end goal under the existing network state in terms of the desired end-to-end network performance.
5. *Plan Model*: Considering the end-to-end performance as the goal, it is used to establish the model of the plan, guiding for the decision-making.
6. *Decision Model*: Using the plan as a guide, it is used to establish the optimal decision-making, and guide for the implementation of the reconstruction.
7. *Reconfiguration Model*: It is used to establish the reconstruction model for the implementation of the reconstruction command in terms of the optimal decision making.

The novel theoretical architecture is driven by the end-to-end goal and achieves the function of learning by the continuous interaction between the module and the feedback of the module. The cognitive information of the theoretical model flows from the Information Model and Projection Model to End-to-end Goal, Plan Model, Decision Model and Reconfiguration Model, realizes the cognitive information rational and orderly flow, namely cognitive flow, and reflects the fundamental role of cognitive information in the entire CWN operation. It's the driving factor of network operation and the important reference factor of optimizing the decision-making.

Compared with the existing architectures of CWN, the advantages of this theoretical model of the cognitive network architecture are:

1. The driven goal based on End-to-end performance
2. The behavior of the cognitive network targeted
3. It fully reflects supporting role of the cognitive flow
4. The relationship between various models in the structure of the theoretical model is clear.
5. The termination condition of the theoretical model is specific (i.e. end-to-end goal)

3.2   Cognitive Information Flow and Resource Flow

3.2.1   *Cognitive Information Flow*

The cognition information is defined as the states of heterogeneous network and the way how the states change, which is achieved by using the information cognition technologies. The function of "cognition" covers all the stages in the cycle, such as sensing, learning, reasoning, planning, deciding, etc. *Cognition flow* is formally defined as the dynamic process of generating, representing, transmitting, and utilizing the cognition information in CWN. Therein, *flow* is used to emphasize the time and network continuity of cognition information. Specifically, from the aspect of time domain, cognition information is time-varying and continuous, while from the aspect of network, cognition information is successively transmitted through CWN, which may span a number of heterogeneous networks. Thus, *cognition flow* is capable of describing the always-changing yet interdependent characteristics of cognition information over time and network domains [12].

The functional architecture is proposed and shown in Fig. 3.4, to improve the spectrum efficiency and the convergence of heterogeneous networks. The function of each module is described as follows.

Cognition Information Management module (CIM) can realize acquisition, preprocessing and unified representation of cognition information.

Intelligent learning Management module (ILM) implements two functions: a) learning/reasoning common cognition information, i.e., the information used by multiple function units when making decisions, which is stored in the Cognition Information/Knowledge Bases unit(CIKB); b) instructing other units to implement intelligent decisions by adopting the knowledge or case-based learning/reasoning mechanism.

CIKB stores cognition information and provides the data query and updates for other function modules. In order to speed up the extraction and storage, CIKB classifies the stored data according to the source, the scope, time granularity or other criteria to improve the distinguishing degree of cognition information.

Advanced Spectrum Management unit (ASM) implements the spectrum management of heterogeneous networks, while Dynamic Network Planning Management & Self-organizing Network unit (DNPM&SON), and Joint Radio

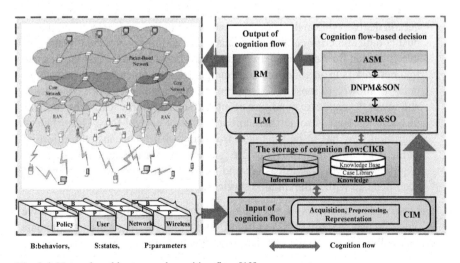

Fig. 3.4 Network architecture and cognition flow [12]

Resource Management &Self-Optimizing unit (JRRM&SO) are responsible for the dynamic planning, self-organization, self-optimizing, and joint radio resource management of heterogeneous networks.

Reconstruction Management unit (RM), by acquiring the network reconfiguration information of DNPM&SON and JRRM&SO, is responsible for initiating the reconfiguration process. RM unit includes Network Reconfiguration Management unit (N-RM) and Terminal Reconfiguration Management unit (T-RM). Therein, N-RM is used to support the reconfiguration of networks, which abstracting and packaging the common functions of multiple heterogeneous networks to support the unified access of mobile users; T-RM locates in the reconfigurable mobile terminals. Apart from the management of reconfiguration information, the setup of reconfiguration software and the implementation of decisions, it also realizes other reconfiguration processes, e.g., network detection, negotiation and software downloading, by interacting with the N-RM in the network side.

Cognition flow originates from the states of four domains, and is used by numerous functional modules across the heterogeneous network. This leads to the key features of cognition flow which inevitably impact the design of network architecture and are identified as follows.

1. *Diversity*: Cognition flow is composed of the state information from four domains, and in each domain, the state information is various and differs from each other. Therefore, for a certain network element at a particular time, the information carried by cognition flow belongs to a subset of all the state information in four domains. That is, the carried information may vary from one cognition flow to another, which is referred to as the *diversity* of cognition flow. For example, a cognition flow may embody the radio parameters at physical layer, the service conditions at data link layer, or the network topology at network layer.

   • This feature requires that cognition flow can be transmitted with the diverse QoS provisioning. For insurance, the sensed raw data may suffer errors to a certain degree, while a decision instruction must be error-free by using the robust coding and error control mechanisms. In addition, occasionally, the data in a cognition flow must be delivered in time, possibly due to that the destination network is in an abnormal and emergent state. Sometimes, a controllable delay is allowed to loosen the requirements for an inter-network connection, and lower the capital expenditure of infrastructures. Intrinsically, the function of cognition flow is somewhat similar to that of control flow. Nevertheless, cognition flow has to be separated from control flow because the priorities of the conventional control flow are guaranteed with the most stringent QoS requirements, which is infeasible and inefficient for diverse cognition flow, accompanied with differentiated services.

2. *Dynamics*: The dynamics of cognition flow are twofold: from the perspective of time, the data amount of a cognition flow is time-varying, and may fluctuate dramatically. It is due to the fact that the contents delivered by a cognition flow

may vary dramatically over a relatively small time scale; from the perspective of network, the amount of the cognition flow between two specific networks also change irregularly.

- This feature indicates that a flexible mechanism is demanded to map the data in cognition flow into logic channels with an elastic bandwidth, which renders another reason for partitioning cognition flow from control flow, since the mechanism is typically unavailable in the realization of control flow transmission.

3. *Heterogeneity*: Cognition flow is usually transmitted among heterogeneous networks, which means that the lower-layer bearers to support cognition flow transmission may be of great difference owing to the differentials of protocols and specifications. In this sense, cognition flow is heterogeneous from network to network.

- From the angle of inter-operation, this feature compels the unified representation of cognition flow to facilitate the understandings of cognition information among heterogeneous networks.

4. *Hetero-sink*: A sink is defined as the destination of a cognition flow. In general, a cognition flow can be used to support or notify a decision timely, or is stored for future utilization. Accordingly, the sink of a cognition flow may be the online receivers or a database.

- This feature impacts on how a cognition flow is transmitted. If only one receiver is demanding the data of the cognition flow, then Cognitive Dedicated Channel (CDC), e.g. single-point mode, is selected for transmission. On the contrary, if many receivers are waiting for the data, then Cognitive Pilot Channel (CPC), e.g. multiple-point mode, is presented to broadcast the common information to save the channel resources. The concept and implementation of CPC are the main contributions to ITU-R [13]. In the meantime, a cognitive database is also necessary to save cognition information efficiently, which enables the smart evolution to CWN.

Currently, different operators are granted different licensed spectrums and wireless system standards, which leads to the co-existence of multiple radio access technologies (RATs) that may affect each other or sometimes induces excessive interference. In China, there are many RATs that are able to support user access, such as GSM, TD-SCDMA, WCDMA, CDMA-2000, WLAN, TD-LTE and so on. Each network plans its own coverage area independently, and the phenomenon of coverage holes and excessive coverage coexists. Due to this fact, networks using adjacent spectrums interfere with each other, which seriously harms the overall utilization efficiency of heterogeneous networks and the end user experience.

To solve this problem, the techniques of cognitive radio are utilized to improve the coordination among networks and promote the convergence of heterogeneous networks, which can provide more favorable services to users. Cognition flow, acting as the sole interaction means of heterogeneous networks, plays an important role in the convergence of heterogeneous networks.

There are two ways through which cognition flow promotes the convergence of heterogeneous networks:

1. Perfection: It is also called network integration. Considering the compatibility, each heterogeneous network still works under its own networking mode and protocol standards. When cross-network service appears, cognition flow notifies the related networks about user states in advance, such as service type, spectrum, channel and so on. Then cognition flow coordinates different networks to provide successive services to users, forming the mechanism which is similar to virtual routing among heterogeneous networks as shown in Fig. 3.5. The users are not aware of the handover among different heterogeneous networks, and network services are transparent to them.
2. Revolution: It is also called network convergence. Different heterogeneous networks are reconstructed to form a unified networking mode and work under the same protocol standards based on user requirements. Therein, cognition flow needs to collect state parameters of all heterogeneous networks involved, such as network standards, occupied spectrums, ongoing services and so on. These parameters are used for reconstruction entity to make decisions, which are subsequently informed to each heterogeneous network by cognition flow. As shown

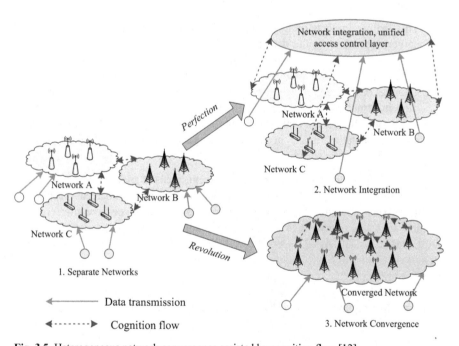

Fig. 3.5 Heterogeneous network convergence assisted by cognition flow [12]

in Fig. 3.5, the networks perform reconstruction in time based on these decisions and then provide the identical access for users.

In order to clarify the advantages and drawbacks between Perfection and Revolution, the comprehensive comparisons, including complexity, compatibility, efficiency and other issues, are made in detail, which are similar to the contrasts between loosely-coupled interworking and tightly-coupled interworking.

1. Implementation complexity: Perfection can be implemented with a lower complexity, since only a unified access control layer is demanded, and each heterogeneous network is actually deployed and organized independently with an interface to the control layer for cognition information interaction. For Revolution, RM is necessitated by each heterogeneous network, which unavoidably compels the equipment updated for legacy networks, and leads to the more complicated realizations for future networks. In addition, RM in each heterogeneous network must work cooperatively to find out the best networking mode and protocol standard for reconstruction, which further increases the implementation complexity greatly.
2. Backward compatibility: Perfection possesses a better backward compatibility than Revolution, for the former almost requires no change inside each heterogeneous network, while the latter acts only if RM is available for heterogeneous networks. Unfortunately, this condition cannot be satisfied for most of legacy networks, which means that substantial network update is necessary to enhance the reconstruction ability.
3. Interworking efficiency: For Revolution, a unified network with the identical networking mode and protocol standard is eventually formed to serve the users originally in different heterogeneous networks. Thus, its interworking efficiency is considerably higher than that of Perfection, in which heterogeneous networks operate individually, and are coordinated by the unified control layer. Potential shortcomings include connection deterioration or even drop during handover, an overlong end-to-end delay owing to the excessive coordination among networks, the poor performance of supporting real-time traffic, etc.
4. Transmission of cognition flow: For Perfection, the transmission of cognition flow may be active all the time, as the coordination among networks is always needed to maintain the ongoing connection stability across multiple networks. However, the situation is quite different for Revolution, in which massive cognition flows are basically transmitted to contribute to the network convergence, and the amount of cognition flow may decrease dramatically once a unified network is constructed, due to the fact that the coordination is relatively simple within the unified network.
5. Challenges for network operators: The current authentication and billing mechanisms are suitable for Perfection, because separate procedures are adopted by each heterogeneous network. As to Revolution, the authentication and billing will encounter serious troubles especially when networks are owned by different operators. For example, after users are served by a unified network, how to balance the revenue among operators? This may be in the scope of national policy,

rather than a technical solution. Besides, since RM must be implemented in each heterogeneous network for the former, the capital investment of Revolution is much larger than that of Perfection for network and equipment updates.

3.2.2   Resource Flow

To address the requirements of the development of future wireless networks, the concept of resource flow is proposed to define the features of resource representation and utilization for the theory of wireless resource knowledge and science. Furthermore, this concept can assist in more advanced unified CRM approaches to guarantee the end-to-end performance. Motivated by that of traffic flows (TFs) in [5, 9, 10], the resource flow (RF) was defined as "A traffic flow is a logic group of packets which have a common attribute. This attribute may be the QoS class or the application the packets belong to".

Definition 1 Resource flow:It is a logical unidirectional flow of resource locomotion, transfer, and transformation of different varieties/dimensions of resources in a wireless communication system.

Definition 2 Streaming of resource flow:It indicates the resource locomotion, transfer, and transformation of multi-dimensional resources motivated by the requests of traffic flows and other effects of resource dissipation and accumulation in communication network ecological systems.

Resource flow is the high-level abstractions of different varieties/dimensions of resources encountered by any traffic flows in a wireless communication system. It includes both a description of the resource itself and the process of resource locomotion, transfer, and transformation. Moreover, resource flow describes the resource in the context of the radio (R), network (N), computing (C), and device (D) environment, and the streaming of resource flow describes the process of resource locomotion, transfer, and transformation.

3.2.2.1   Scientific Connotations

Resource flow provides a new concept and a train of thought for better achieving wireless resource knowledge and effectively utilizing the wireless resource. The concept of resource flow embodies a variety of scientific connotations, including the complexity, dynamics, and temporal and spatial characteristics, as illustrated in Fig. 3.7. The complexity, dynamics, as well as temporal and spatial characteristics are introduced below.

1. Complexity refers to the multi-dimensional complicated context of the user, radio and network environment; the various kinds of resources; the flexible CRM scheme and the random resource requirements of the different users. In CWNs, all the possible environments in which a cognitive terminal could be located and

Fig. 3.6 Multi-agent-based cognitive resource management (MA-CRM) framework

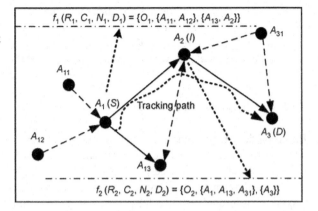

all the related resource types managed by a flexible control scheme are observed to maintain the various QoS requirements of different users with a guarantee of end-to-end performance. Therefore, resource flow provides a unified penetration framework for different environments, resources, schemes, and users.

2. Dynamics refer to the dynamic changes for the resource utilization in the multi-dimension environment; the mobility of various types of resource; and the user traffic switch-over property. Resource flow is a systematically dynamic process, which is closely bounded to resource demands, the environment, and pricing decisions of the operators. On the other hand, resource flow is continuously being updated with respect to format and quantity according to the resource dissipation and storage.

3. Temporal and spatial characteristics are the basic characteristics of resource locomotion, transfer, and transformation. Resource flow has properties of both temporal and spatial characteristics, which are manifest in two ways. First, implement the resource flow attributes with different regular patterns and behaviors due to

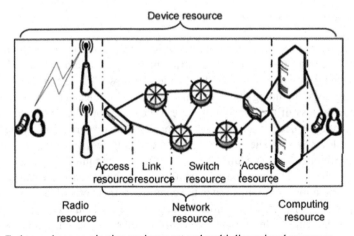

Fig. 3.7 End-to-end communication environment and multi-dimensional resources

the various types of resources and different states. Meanwhile, the same resource located in different environments would cause resource flow to exhibit different behaviors. Second, resource flow runs in various spaces including inter-network, intra-network, and the cross-layer and intra-layer of the communication protocol.

3.2.2.2   Classification of Resource Flow

Based on their ranges and the subjects upon which they act, resource flow is classified as the longitudinal resource flow and latitudinal resource flow. A latitudinal resource flow is the resource locomotion, transfer, and transformation of a multi-dimensional resource crossing different networks (heterogeneous and homogeneous). A latitudinal resource flow runs from a resource-rich network to a resource-poor network, which indirectly achieves the load balancing performance. For example, in LTE networks, each enhanced NodeB (eNB) should implement the load balancing in a self-organizing way to balance the traffic load; however, the use of the latitudinal resource flow can avoid complicated manipulation and implementation.

A longitudinal resource flow is the resource locomotion, transfer, and transformation of a multi-dimensional resource crossing different protocol layers in the same communication protocol stack, which is mostly reflected by the resource variance in function and worth in different layers. Research on the longitudinal resource flow will help to improve the resource utilization in different layers and further achieve a better overall network performance. Various types of resources are in the protocol stack, including physical resources, such as power, media access resources, such as switches, and network resources, such as routers. The resource possesses different functional properties, including power, delay, and error correction. Therefore, the longitudinal resource flow with typical functional properties can satisfy and balance resource requests and resource mapping in the protocol stack.

3.2.2.3   Characterization of Resource Flow

To characterize the resource flow, the definition of the electric current is recalled firstly. If there is an electric potential difference in a conductor, there will be an electric current running through it. Meanwhile, the flow of the electric current will cause energy loss, including the electrical energy loss of the electric resistance and thermal energy loss of the conducting wire. Likewise, both the resource potential and resource resistance are proposed to characterize the resource flow in the "communication logic wire". The resource flow is certainly motivated by the resource potential difference between different agents in the communication scenario in Fig. 3.8, which contains the resource potential and resource resistance caused by resource accumulation and dissipation in the communication network eco-logical systems, respectively. Resource potential is the resource accumulation, which means that if the agent can obtain more resources from others, the agent will be at a high

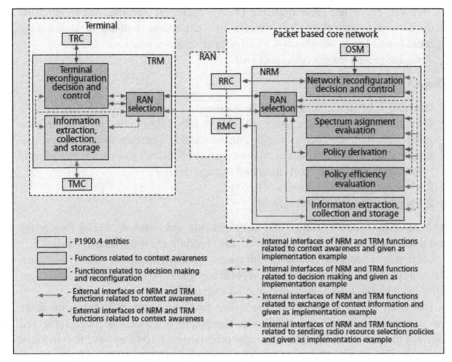

Fig. 3.8 The P1900.4 functional architecture [2]

resource potential level. For example, in Fig. 3.6, the resource supporters A_{11} and A_{12} of agent A_1 enhance the resource potential of A_1. Essentially, resource resistance is resource dissipation, which is the result of the resource demand of the traffic flows and the resource expenditure to other agents. In Fig. 3.6, A_{13} and A_2 act as the resource expenditure of A_1, which is one part of the resource resistance, and the other part is the resource demand of the traffic flows towards agent A_1. The effects of the resource resistance are taken by the resource potential. Furthermore, it is assumed that resource potential is a vector, which has a direction; on the other hand, resource resistance does not have a direction and is scalar.

3.3   Functional Architecture Model

The IEEE 1900.4 [2] Working Group has proposed the functional architecture defined in the current version of the draft standard. This functional architecture is shown in Fig. 3.8. The whole architecture is divided into two parts: Terminal part and Packet based core network part.

The terminal side modules:

1. Terminal reconfiguration controller (TRC)

2. Terminal reconfiguration manager (TRM) includes: Terminal reconfiguration decision and control module, Information extraction, collection, and storage module and Radio access network (RAN) selection module.
3. Terminal measurement collector (TMC)

The network side modules:

1. Operator spectrum manager (OSM)
2. Network reconfiguration manager (NRM) includes: RAN selection module, Network reconfiguration decision and control module, Spectrum assignment evaluation module, Policy derivation module, Policy efficiency evaluation module and Information extraction, collection and storage module.
3. RAN reconfiguration controller (RRC)
4. RAN measurement collector (RMC)

RAN selection module connects the Terminal side and Network side by the peripheral interface. The interface functions of each module are illustrated in Table 3.1.

Moreover, European Telecommunications Standards Institute (ETSI) also proposed a functional architecture for CWN mainly focusing on reconfiguration capability [8] which is based on cognitive pilot channel [4].

However, the architectures in the above two researches only focus on some specific applications. They just focus on the cognitive process of single-link, and have not formed the CWN. Moreover, the proposed architectures have four models: Cognition flow model, Function model, Database model and Transmitter management, which are not enough to guarantee the end-to-end performance of the CWN from a network perspective. The novel functional architecture is the application

Table 3.1 Interface function

Interface	Direction of connection	Function
Interface between the NRM and TRM	From NRM to TRM	Radio resource selection policies
		RAN context information
		Terminal context information
	From TRM to NRM	Terminal context information
Interface between the NRM and RRC	From NRM to RRC	RAN reconfiguration requests
	From RRC and NRM	RAN reconfiguration responses
Interface between the NRM and RMC	From NRM to RMC	RAN context information requests
	From RMC to NRM	RAN context information
Interface between the TRM and TMC	From TRM to TMC	Terminal context information requests
	From TMC to TRM	Terminal context information
Interface between the TRM and TRC	From TRM to TRC	Terminal reconfiguration requests
	From TRC to TRM	Terminal reconfiguration responses
Interface between the NRM and OSM	From OSM to NRM	Spectrum assignment policies
	From NRM to OSM	Information on spectrum assignment decisions

of the theoretical model of the wireless network architecture based on the cellular network. Based on the theoretical model structure, the functional structure needs to implement the following functional requirements:

1. The functions of acquisition, processing, characterization and transfer the multi-domain cognitive information.
2. End-to-end performance as the goal of the learning function.
3. The ability of end-to-end performance as the goal, autonomous optimization of multi-domain resource management.
4. The ability of reconfiguration.

The functional architecture of the CWN includes the network side and terminal side in Fig. 3.9. The network side and terminal side consist of four functional modules:

1. Cognitive Information Management (CIM), mainly used for the acquisition, processing, characterization, and delivery management of cognitive information.
2. Database & Intelligence Management (DB&IM), mainly used for learning and storing long-term cognitive information and knowledge.
3. Network Convergence Management (NCM), mainly used for self-optimizing management of multi-domain resource management as the end-to-end goal.
4. Reconfiguration Management (RM), mainly used for the management of CWN reconstruction decision-making and implementation.

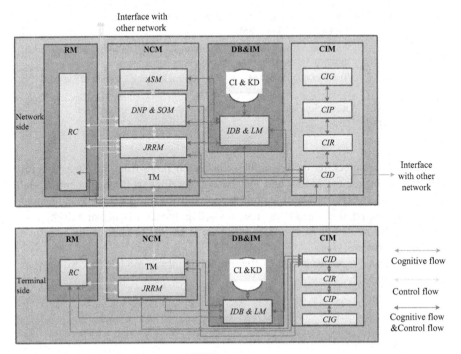

Fig. 3.9 The functional architecture of CWN

Table 3.2 The function of modules

Management model	Sub-model	Function
Cognitive information management (CIM)	Cognitive information gathering (GIG)	Mainly used to obtain cognitive information from multi-domain environment
	Cognitive information processing (CIP)	Pretreatment processing the cognitive information obtained from multi-domain environment, and translating into meaningful information of network
	Cognitive information representation (CIR)	Characterized the pretreatment cognitive information to objectively reflect the true state of the multi-domain environments
	Cognitive information delivering (CID)	Transmitted the characterized cognitive information
Database & intelligence management (D&IM)	Cognitive information & knowledge database (CI%KD)	Store long-term cognitive information and knowledge for better implementation of network integration and reconstruction optimization decisions
	Intelligent database & learning management (ID&LM)	Manage the operation of cognitive information & knowledge database module and realize the learning function
Network convergence management (NCM)	Advanced spectrum management (ASM)	Manage long-term spectrum utilization policy. Long-term spectrum management strategy could interact between the different cognitive wireless networks by the module
	Dynamic network planning & self-organizing management (DNP&SM)	Planning and management of the network dynamic and autonomous
	Joint radio resource management (JRRM)	Coordinated manage multiple network radio resources
	Transfer manager (TM)	The control and management of the transmission signal
Reconfiguration management (RM)	Reconfiguration control (RC)	Control the network and terminal reconstruction

In this CWN architecture, the cognitive information flows from the Cognitive Information Management Model and Database & Intelligence Management Model to Network Convergence Management Model and Reconfiguration Management Model, to support these modules' corresponding function. Control information transmitted by related cognitive flow interface is mainly to control the cognitive information transmission.

The functional modules include different functional sub-modules accordingly, as shown in Table 3.2.

3.4   Deployment Architecture Model

In terms of the heterogeneous networks with various coverage types such as macro, pico and femto cells, the density of user demands is unevenly distributed around a large area in Fig. 3.10. In order to improve the network capacity of the proposed architecture, a layered heterogeneous network architecture is designed and proposed in Fig. 3.10 based on the separation framework of cognitive information, control information and data information. The deployment architecture can effectively improve the efficiency of the resource utilization and reduce the network control overhead.

To quantify the efficiency of resource utilization, new resource utilization metrics are proposed and defined in this section. Considering the contribution of space-time-frequency three-dimensional resources to the network capacity, as well as transmission ratio of useful information by users, the space-time-frequency radio resource efficiency is denoted in Eq. (3.1).

$$\eta_e = \frac{\alpha C_0}{L_2[S]L_1[T]L_1[W]} \tag{3.1}$$

The definition considers about the network coverage S, time T, the system bandwidth occupied W, the transmission capacity C_0, as well as useful information α, L (*) as a measure function. In the actual network, users are unevenly distributed in a large area which is dynamically changing over a period of time. The uniformly distributed network with predefined and planned coverage can not effectively match the variety of users in temporal and spatial domains, leading to the low resource utilization. Therefore, both the new network architecture and network deployment technologies are proposed under the case of unevenly distributed user

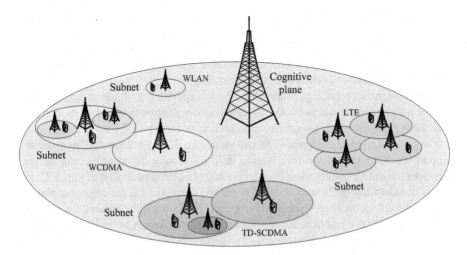

Fig. 3.10 Layered Heterogeneous Network Model

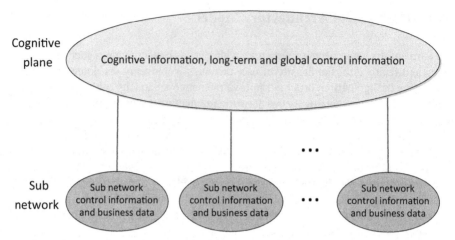

Cognitive plane

Cognitive information, long-term and global control information

Sub network

Sub network control information and business data

Sub network control information and business data

• • •

Sub network control information and business data

Fig. 3.11 The separation model of control information, business and cognitive information

demands, achieving the exact match between user demands and the resources. The new network architecture has the following characteristics:

1. Match the distributions of network and user information density, and the overall is a heterogeneous hierarchical network.
2. The small cell and macro cell co-exist: the macro cell covers the entire region and the small cell is used to increase the capacity.
3. The coexistence of a variety of communication systems: different access technologies complement each other, and optimize the efficiency of resource utilization.

By analyzing the technologies and features of a variety of networks, the control signaling and data transmission of the current wireless network are overlapped in coverage. And this overlap has two problems. First, in order to achieve the heterogeneous networks integration, some control information must be extracted from each network for a unified scheduling. Second, configuring each network with a set of control channels will lead to underutilization of the network control channel. Therefore, a large coverage network for control signaling delivery is proposed to improve the utilization of the control channel resources.

Therefore, new technologies are proposed to separate the global and long-term control information about network integration from each network, and then combine these information and cognitive information to form a Cognitive Plane which is a cognitive-based control plane, as shown in Fig. 3.11. Furthermore, in order to support the cooperation and information delivery among different heterogeneous networks, cognitive information flow technology has been proposed in Fig. 3.12 to enable the cognitive information delivery among networks via the cognitive plane, which is based on the theory proposed in Chap. 2. Besides, the resource flow is applied to indicate the resource locomotion and transformation in multi-dimensions by the traffic requests in heterogeneous networks.

Fig. 3.12 Architecture of
heterogeneous network based
on cognitive level

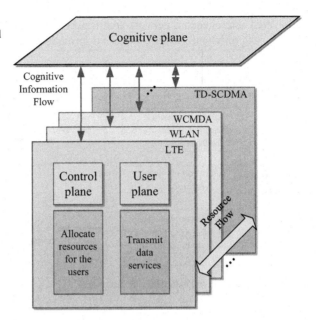

In this architecture, a layered heterogeneous network architecture is proposed
with the separation framework of cognitive information, control information and
data information. The network is divided into two layers: the Cognitive Plane and
the Sub Network.

The Cognitive Plane is mainly used to manage the cognitive information, as well
as a small amount of long-term and global control information of networks, such
as establishing a connection for terminals, selecting transmission network for user
services, hand over between networks and so on.

The Sub Network includes a variety of different types of heterogeneous networks. Each Sub Network is responsible for the management of its internal control
information and the user's data service.

The core of the architecture is the separation of the cognitive information, control information and data information. Cognitive information is managed uniformly
by the Cognitive Plane.

Then, a brief introduction of the working process of the Cognitive Plane and the
Sub Network is proposed and shown in Fig. 3.13.

First, when users turn on a terminal, it will access to the Cognitive Plane according to the broadcast information of the Cognitive Plane.

Second, after accessing to the Cognitive Plane, the terminal detects the signal
intensity of the available Sub Network, and sends the network ID number, received
power, degree of interference, as well as terminal ID and movement speed to the
Cognitive Plane periodically.

Third, when the terminal needs to initiate a service, it sends a request to the
Cognitive Plane. Cognitive Plane will collect the information provided by the user
before, such as the type of service, the load and resource usage of related networks.

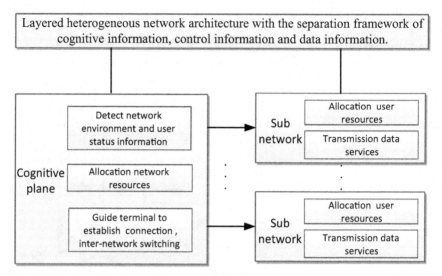

Fig. 3.13 The Cognitive Plane with Sub Networks

Then the Cognitive Plane selects the appropriate network as the serving cell to maximize the network capacity.

Finally, the Cognitive Plane sends the information of the terminal to the serving cell, which will be responsible for the data transmission.

Compared with the existing technical solutions, this architecture reflects a new way for future network deployment: pursuing network capacity and high data rate at hot spots instead of network coverage. The Sub Network is applied to guarantee the high data rate in the hot spots area, while ensuring that the terminal is always online to guarantee the quality of service by controlling the Cognitive Plane coverage.

3.5 Concluding Remarks

In this chapter, a novel architecture of CWN is proposed, including the theoretical architecture and the functional architecture. And both the cognitive information flow and resource flow are proposed to depict the information flows of cognitive information and resource information. Furthermore, the functional architecture and deployment architecture models are also described in this chapter, respectively.

References

1. Allen BM, Jeffrey HR, Peter A (2009) Cognitive radio and networking research at Virginia Tech. Proc IEEE 97(4):660–689

2. Buljore S, Harada H, Filin S et al (2009) Architecture and enablers for optimized radio resource usage in heterogeneous wireless access networks: the IEEE 1900.4 working group. IEEE Commun Mag 47(1):122–129
3. Denkovski D, Pavlovska V, Atanasovski V et al (2010) Novel policy reasoning architecture for cognitive radio environments. IEEE GLOBECOM, Miami, 6–10 Dec 2010
4. Feng ZY, Zhang QX, Tian F et al (2012) Novel research on cognitive pilot channel in cognitive wireless network. Wireless Pers Commun 62(2):455–478
5. Mark AM, Peter AT, Dan M (2006) Chicago spectrum occupancy measurements & analysis and a long-term studies proposal. Paper presented at the Proceedings of the first international workshop on Technology and policy for accessing spectrum, New York, 2006
6. Mitola J (1993) Software radios: survey, critical evaluation and future directions. IEEE Aero El Sys Mag 8(4): 25–36
7. Mitola J (1999) Cognitive radio: making software radios more personal. IEEE Pers Commun 6(4):13–18
8. Mueck M, Piipponen A, Kalliojafirvi K et al (2010) Etsi reconfigurable radio systems: status and future directions on software defined radio and cognitive radio standards. IEEE Commun Mag 48(9):78–86
9. Simon H (2005) Cognitive radio: brain-empowered wireless communications. IEEE J Sel Areas Commun 23(2):201–220
10. Spectrum Dashboard (2010) REBOOT FCC GOV, Washington DC, http://reboot.fcc.gov/spectrumdashboard/searchSpectrum.seam. Accessed 17 Nov 2010
11. Thomas R, Friend D, DaSilva L et al (2006) Cognitive networks: adaptation and learning to achieve end-to-end performance objectives. IEEE Commun Mag 44(12):51–57
12. Xu WJ, Lin JR, Feng ZY et al (2013) Cognition flow in cognitive radio networks. China Commun 10(10):74–90
13. Zhang P, Liu Y, Feng ZY et al (2012) Intelligent and efficient development of wireless networks: a review of cognitive radio networks. Chin Sci Bull 57(28–29):3662–3676

Chapter 4
Cognitive Information Awareness and Delivery

According to the analysis above, cognitive information awareness is the first step for CWNs to gather the necessary network information such as the available spectrum and network operation parameters. Cognitive ability can be mainly categorized as spectrum sensing, cognitive pilot channel (CPC) and cognitive database according to different information and collection methods. Wherein, spectrum sensing determines the available spectrum parameters including frequency, bandwidth, and idle period. The CPC and cognition database can be used to collect and exchange the network information such as the radio access technology (RAT) mode, network pilot channel information, system bandwidth, carrier frequency, transmit power, and policies.

4.1 Intelligent Spectrum Sensing

Spectrum sensing is defined as the task of detecting spectrum holes by sensing the radio spectrum in the local neighborhood of the CWN in an unsupervised manner [1]. Spectrum holes, i.e. spectrum opportunities, are defined as "a band of frequencies that are not occupied by the primary user of that band at a particular time in a particular geographic area" [2]. The spectrum holes can be modeled in the dimensions of frequency, time and space. Furthermore, there are other dimensions for spectrum holes, such as the code and angle [3]. Thus, the objective of spectrum sensing focuses mainly on the multi-dimensional properties of spectrum for exploring more spectrum holes.

To be specific, the task of spectrum sensing can be summarized as follows.

1. Detection of spectrum holes: spectrum sensing detects the spectrum holes, and models them according to the multi-dimensional properties based on the requirements of reliable communication for secondary users;
2. Interference analysis and signal classification: a signal is detected and analyzed for classification according to its characteristics. The interference can be distin-

© The Author(s) 2015
Z. Feng et al., *Cognitive Wireless Networks,* SpringerBriefs in Electrical
and Computer Engineering, DOI 10.1007/978-3-319-15768-9_4

guished from signals via characteristic analysis so that the CWN can be reconfigured to avoid harmful interference.

To achieve these tasks above, numerous technologies are proposed to conduct spectrum sensing which include local detection and cooperative detection.

4.1.1   Local Detection

This focuses on the detection of the existence of the primary user's signal based on signal processing methods. The sensing node (secondary base station or users) analyzes the received primary user's signal, and decides whether primary users are present or not. The challenge is how to limit the transmit power of secondary devices to avoid the interference to primary users. Thus, the FCC quantifies the measurement metric as the "interference temperature" for interference assessment [4]. It is proposed to maximize the amount of tolerable interference for a given spectrum band at a particular location. The secondary users are required to transmit under the constraint that the sum of the existing noise and interference does not exceed the interference temperature threshold at a licensed receiver or primary user. The three main categories of local detection are energy detection, feature detection and matched filter detection.

Energy detection is the most common method for spectrum sensing due to its low implementation complexity. It requires no prior knowledge of the primary user's signal, and does not need a special design for detecting spread spectrum signals [5]. Due to the comparison between the output of the energy detector and the given threshold for primary users, the challenges for energy detection are the selection of the appropriate detection threshold, the inability to differentiate interference from primary users and noise, and the poor performance under low signal-to-noise ratio (SNR) values [6]. There has been abundant researches to address these challenges, such as [7, 8]. The typical version of energy detector is shown in Fig. 4.1.

Feature detection derives from the specific features associating with the modulated signals transmitted by primary users. In many cases, signals have periodic statistic features such as modulation rate and carrier frequency which are usually

Fig. 4.1 The typical version of energy detector

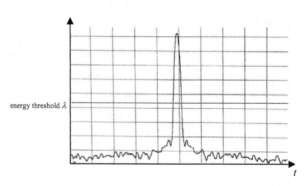

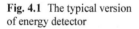
energy threshold λ

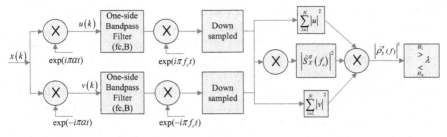

Fig. 4.2 Discrete-time version of cyclostationary feature detector [9]

viewed as cyclostationary characteristics. In detection, the cyclostationary characteristic of a primary user's signal can be distinguished from noise in its statistical
properties such as its mean and autocorrelation values [9, 10]. Compared with energy detection, cyclostationary detection is not sensitive to the noise uncertainty, so
it is more robust in low SNR regimes. However, this method requires more prior
information on the primary user's signals for deciding the occupancy of primary
user, which leads to a much greater complexity [11, 12]. The typical structure of
cyclostationary detector is shown in Fig. 4.2.

Matched filter detection is an optimized detection method based on the primary user's signal as prior knowledge for the secondary users. The advantage of the
matched filter detection is the short time to achieve a lower probability of a false
alarm and a lower probability of a misdetection. The matched filter method requires
fewer signal samples, which decreases as a target probability of false alarm at low
SNRs [13, 14]. Thus, there exists a SNR wall for a matched filtering method [15].
Moreover, matched filter detection requires that the received signals are demodu-
lated. Perfect knowledge of the primary user's signal is required, so the implementation complexity of a sensing unit is impractically high [16]. To solve the SNR
wall problem, waveform-based detection is proposed to improve the match filtering
detection when the number of samples is large enough [17]. However, information
on the patterns of the primary user's signal is a prerequisite for implementing the
waveform-based detection, so minimizing the implementation complexity is still an
open issue.

4.1.2   Cooperative Sensing

The performance of spectrum sensing may be deteriorated due to the noise uncertainty, shadowing, multi-path fading and the SNR wall, especially in the problem of
the hidden primary user. Therefore, the cooperative sensing is proposed as an efficient solution to solve these issues. Furthermore, cooperative sensing can efficiently
solve the problem of the hidden primary user in an efficient way [18]. But there
are also challenges, including the network architecture of centralized or distributed cooperation sensing [19, 20], detection fusion including decision fusion or data
fusion [21, 22] and cooperative node selection. In order to solve these problems,

researches in [23, 24] have improved the cooperation sensing performance from the space diversity and the abnormality detection perspectives.

Coordinated spectrum sensing ensures cognitive radio systems to get the usability of each channel within the entire spectrum of interest by assigning terminals to detect different channels simultaneously. [25] has proposed an optimized channel-assignment strategy for coordinated spectrum sensing based on an Iterative Hungarian Algorithm. Simulation results show that the proposed strategy performs well for improving the overall sensing performance of the system.

In coordinated spectrum sensing, the entire spectrum of interest is divided into several narrow channels, which can be detected by each terminal with a decent accuracy during each sensing slot [26]. Then each terminal is assigned to detect one channel with the precondition that each channel is detected by at least one terminal if possible. To address the optimization of the channel assignment strategy in coordinated spectrum sensing, each terminal is assigned to detect a channel in consequence of its sensing performance on the respective channels to achieve an optimized overall sensing performance. The problem can be formulated as a modified Linear Assignment Problem (LAP), and an Iterative Hungarian Algorithm is proposed to find the optimized solution.

In the system, false-alarm probability P_{fa} is set to be identical for all the local sensing so that misdetection probability P_{md} could alone indicate how reliable the sensing is. The sensing performance matrix can be defined as Eq. (4.1).

$$P_{md} = \left[\left(P_{md} \right)^{n,m} \right]_{N \times M} \tag{4.1}$$

where $n = 1, 2, \dots, N$, $m = 1, 2, \dots, M$. For each terminal, the spectrum sensing performance of different channels $(P_{md})_i$ usually varies. To achieve the best overall sensing performance in coordinated sensing, an efficient channel-assignment strategy based on P_{md} should be applied.

Given a unified local P_{fa} and a universal performance requirement of the system $Q_{fa} = Q_{mfa}$, $m = 1, 2, \dots, M$, the maximum number of terminals cooperatively sensing one channel is denoted in Eq. (4.2).

$$N_{\max}^{coop} = \left\lfloor \log_{(1-P_{fa})} \left(1 - Q_{fa} \right) \right\rfloor \tag{4.2}$$

The overall sensing performance can be indicated by the sum of probability of misdetection over all channels, which should be minimized in the coordinated sensing strategy. The assignment matrix $X = \left[x^{nm} \right]_{N \times M}$ is introduced for a better expression of the problem in Eq. (4.3).

$$x^{nm} = \begin{cases} 1 & f(n) = f_m \\ 0 & else \end{cases} \tag{4.3}$$

So the optimal coordinated sensing problem is formulated as Eq. (4.4).

$$\min_{X} \sum_{m=1}^{M} Q_{md}^{m} = \sum_{m=1}^{M} \left(\prod_{x^{nm}=1}^{M} (P_{md})^{n,m} \right) \tag{4.4}$$

$$s.t. \sum_{m=1}^{M} x^{nm} = 1, n = 1, 2, ..., N;$$

$$a \le \sum_{m=1}^{M} x^{nm} \le N_{\max}^{coop}, m = 1, 2, ..., M; \tag{4.5}$$

$$x^{nm} \in \{0,1\}$$

where $a = \begin{cases} 1 & N \ge M; \\ 0 & N < M; \end{cases}$

Based on the Iterative Hungarian Algorithm in [27, 28], a detailed description of the algorithm is given as follows.

Step 1: Initialization:

- Input: N, M, P_{md}, P_{fa}, Q_{fa}
- Initialize the operation matrix $P = \left[p^{n,m} \right]_{N \times M}$ as $P = P_{md}$;
- Initialize the cumulative assignment matrix $X = 0$;
- Initialize the iteration assignment matrix $X_0 = 0$;
- Initialize the cumulative misdetection probability vector $\tilde{Q}_{md} = 1$;
- Initialize the number of unassigned terminal $N_{una} = N$.
- Calculate $N_{\max}^{coop}$.

Step 2: Calculate $N_{asgn} = \min(M; N_{una})$ and run the Iterative Hungarian Algorithm for P to assign $N_{\max}^{coop}$ terminals to sense different channels.

$$\min_{X_0} \sum_{n=1}^{N} \sum_{m=1}^{M} p^{nm} (x_0)^{nm} \tag{4.6}$$

Step 3: Append the assignments in this iteration to the overall assignment in Eq. (4.7).

$$X = X + X_0 \tag{4.7}$$

For each terminal assigned to sense a channel in this iteration (denoted as i), update the cumulative misdetection probability of the corresponding channel.

$$\tilde{Q}_{md}^{f(i)} = \tilde{Q}_{md}^{f(i)} (P_{md})^{i,f(i)} \tag{4.8}$$

Step 4: For each unassigned terminal (denoted as j), update the cooperation gain for each channel (denoted as m).

$$p^{jm} = \tilde{Q}^m_{md} - \left(P_{md}\right)^{j,m} \tilde{Q}^m_{md} \qquad (4.9)$$

Step 5: For each assigned terminal (denoted as k), the parameter is set in Eq. (4.10).

$$p^k = 0 \qquad (4.10)$$

Step 6: Provided that there are no unassigned terminals or $N_{max}{}^{coop}$ terminals have been assigned to sense each channel, the algorithm terminates and X denotes the final assignment strategy, or set the parameter in Eq. (4.11).

$$P = 1 - P \qquad (4.11)$$

Update $N_{una} = N_{una} - N_{asgn}$, restore X_0 as $X_0 = 0$, then go to Step 2 to continue the iteration.

The simulation results are shown in Fig. 4.3, exhibiting that the performance of the proposed Iterative Hungarian Algorithm based strategy is obviously prior to that of the greedy based strategy. When P_{md} is 0.45, the average Q_{md} of the Iterative Hungarian Algorithm based strategy is approximately 0.03 lower than that of the greedy based strategy. The advantage gets more remarkable when P_{md} is 0.3, where the average Q_{md} of the Iterative Hungarian Algorithm based strategy is just a quarter of that of the greedy based strategy.

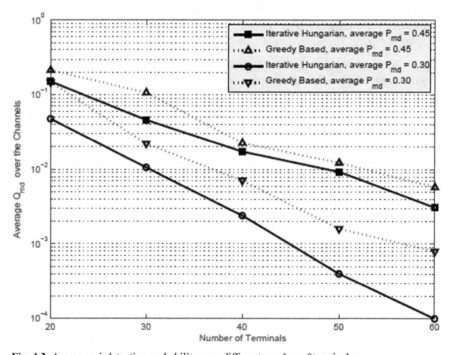

Fig. 4.3 Average misdetection probability over different number of terminals

4.2 Multi-Domain Cognitive Database

Apart from the information of available spectrum, the CWN needs to collect the necessary network information such as operation type, cell coverage, and modulation type. It should also consider new method to store the information on spectrum and network operation, and to exchange the related information to meet user requirements. So, the cognitive databases are proposed to store and manage the network information.

FCC has proposes rules for the secondary unlicensed usage of the TV bands in the United States [29]. ECC (European Communication Committee) published a report in January 2011 to provide technical and operational requirements for the possible operation of cognitive radio systems in the 'white spaces' of the spectrum band 470–790 MHz [30]. Both FCC and ECC have proposed geo-location database usage as the solution for primary system detection and interference avoidance.

Recently, a multi-domain cognitive database was proposed [3, 4]. The network information in the CWN can be classified and organized to improve information management, representation and access.

4.2.1 Geo-Location Database

Geo-location database is a database system that maintains records of all authorized services in the TV spectrum bands. It is capable of determining the available channels at a specific geographic location based on the interference protection requirements. It can also provide lists of available channels to white space devices (WSDs) that have been certified under the regulatory bodies' equipment authorization procedures.

4.2.1.1 Geo-Location Database of FCC

The main idea proposed by FCC is to define a protection radius for the co-channel and the first adjacent channels plus no-talk area, where WSDs are not allowed to transmit.

There are several types of licensed incumbents present in the TV bands in the U.S. today. This includes a variety of different types of analog (NTSC) and digital (ATSC) broadcast operations. Each type of licensed incumbent system has specific interference protection requirements. Each TV station has a commonly regulated protected service area that is determined by its Grade B Contour for analog broadcast operations or its Noise Limited Contour (NLC) for digital broadcast operations. FCC defines the protected service contour levels in terms of a minimum TV signal E-field strength at a nine-meter-high outdoor receiving antenna for various station types. These levels are presented in Table 4.1.

Table 4.1 Criterion for definition of TV station protected contours [29]

Type of station	Protected contour	
	Channel	Contour (dBu)
Analog: class A TV, LPTV, translator and booster	Low VHF (2–6)	47
	High VHF (7–13)	56
	UHF (14–69)	64
Digital: full service TV, class A TV, LPTV, translator and booster	Low VHF (2–6)	28
	High VHF (7–13)	36
	UHF (14–51)	41

Table 4.2 Protection criteria of digital and analog TV receivers [29]

Type of station	Protection ratios	
	Channel separation	D/U ratio (dB)
Analog TV, class A, LPTV, translator and booster	Co-channel	34
	Upper adjacent	−17
	Lower adjacent	−14
Digital TV and class A	Co-channel	23
	Upper adjacent	−26
	Lower adjacent	−28

Whether or not the interference to TV reception occurs depends on the desired-to-undesired (D/U) signal ratio required by acceptable services. FCC defines interference protection criteria that must be applied for both digital and analog TV receivers to avoid interference, which are shown in Table 4.2.

In order to minimize the complexity for compliance, FCC specifies a table of minimum required separation distances between TV station contours and TV WSDs as the application of TV protection criteria. Given the power limit of TV band devices, the height of the device's antenna above ground, and the D/U protection ratios specified above, the separation distances can be calculated in Table 4.3.

WSDs must be located outside the contours of co-channel and adjacent channel stations with the minimum separation distances [30]. Deployment scenarios of co-channel and adjacent channel TV band devices in TV white space are illustrated in

Table 4.3 Minimum required separation distances between unlicensed TV band devices and TV service contour edge [29]

Antenna height of unlicensed device	Required separation (km) from digital or analog TV (full service or low power) protected contour	
	Co-channel(km)	Adjacent channel(km)
Less than 3 m	6.0	0.1
3-less than10 m	8.0	0.1
10–30 m	14.4	0.74

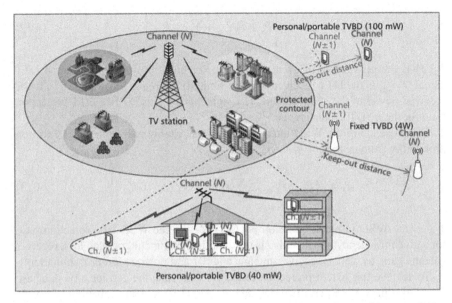

Fig. 4.4 Scenarios of co-channel and adjacent channel TV band devices in TV white space [30]

Fig. 4.4. And the fixed devices are allowed to operate at up to 1 watt (W) transmitter output power and with antenna gains to achieve 4 W equivalent isotropically radiated power (EIRP). Personal/portable operations will be permitted at up to 100 mW EIRP, with no antenna gain, except that when operating on a channel adjacent to a TV station or other licensed station/service and within the protected coverage area of that service, operations will be limited to 40 mW. Therefore, the personal/portable devices can work on the adjacent channel within the protected coverage area under the condition that the power is limited to 40 mW.

4.2.1.2 Geo-Location Database

The Electronic Communications Committee (ECC) aims at specifying the location specific maximum permitted in-block and out-of-block emission limits for WSDs, and to provide protection of ± 10 DTT adjacent channels, apart from the co-channel protection. There are two general operational conditions to be satisfied by a WSD: (1) the device is not allowed to operate within the coverage area of a co-channel BS transmitter; (2) the device may operate within the coverage area of a non co-channel BS transmitter, provided that any reception of this particular BS channel—which could be adjacent to the WSD channel—is protected.

ECC uses the reduction of location probability as the metric for specifying regulatory emission limits for WSDs operating in digital terrestrial TV frequencies. The location probability is defined as the probability that a DTT receiver would operate correctly at a specific location.

$$q_1 = P_r \left\{ P_S \geq P_{S,\min} + \sum_{i=1}^{K} r_{u,k} P_{u,k} \right\} \qquad (4.12)$$

$P_r\{A\}$ is the probability of event A, P_s is the received power of the wanted DTT signal, $P_{s,\min}$ is the DTT receiver's (noise-limited) reference sensitivity, $P_{u,k}$ is the received power of the kth unwanted DTT signal, and $r_{u,k}$ is DTT-to-DTT protection ratio for the kth DTT interference.

The presence of the WSD interference will inevitably reduce the DTT location probability from q_1 to $q_2 = q_1 - \Delta q$.

$$q_2 =_r \left\{ P_S \geq P_{S,\min} + \sum_{i=1}^{K} r_{u,k} P_{u,k} + r(\Delta f) G P_{IB}^{CR} \right\} \qquad (4.13)$$

P_{IB}^{CR} is the WSD in-block emission power, $r(\Delta f)$ is the WSD-BS protection ratio for a given frequency offset, and G is the coupling gain including path loss, receiver antenna gain, as well as receiver antenna angular and polarization discrimination.

To achieve this extent protection of the DTT service, the geo-location database specifies a maximum permitted degradation of DTT location probability, then the location specific maximum permitted in-block and out-of-block emission limits for WSDs can be calculated.

Specifically, for a given geographic pixel, all DTT channels can be potential victims, so the database must examine all relevant co-channel and adjacent-channel interference scenarios with respect to the victim DTT channels. The database must reconcile all calculated WSD in-block and out-of-block emission levels for the given pixel, in order to derive the WSD regulatory emission limits over all DTT frequencies.

As shown in Fig. 4.5, DTT service uses f_1, DTT service uses f_2 and DTT service uses f_3 can be potential victims of CR devices, so the database must examine all the

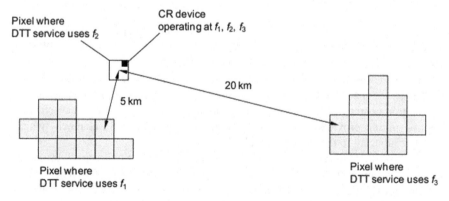

Fig. 4.5 Usage of frequency $f_1, f_2,$ and f_3 by the WSD and the DTT service[31]

cases and derive the appropriate WSD in-block and out-of-block emission levels according to the most critical case.

In conclusion, both FCC and ECC consider that the geo-location database is the most feasible approach considered so far for the protection of incumbent services. However, the approaches FCC and ECC have selected to control the interference generated to the TV incumbent service are somewhat different. FCC specifies fixed transmission power and requires a protection distance around the TV coverage area, while ECC does not specify any protection distance but aims at specifying the location specific maximum permitted in-block and out-of-block emission limits for WSDs.

4.2.2   Multi-Domain Cognition Database

Based on geo-location database, a new database called multi-domain cognition database has been proposed in [32]. The information from the heterogeneous networks is comprehensively related to the network type, operation frequency, location, time slot, user information, and different protocol layers of the networks. Therefore, the information can be classified into several domains accordingly, and the multi-domain cognition database can store and manage them according to the domain divisions, such as the wireless domain, network domain, user domain and policy domain.

1. The wireless domain: The wireless domain consists of the parameters of radio transmission characteristics such as transmit power, spectrum band, Signal to Interference plus Noise Ratio (SINR), transmission rate and the radio resources bandwidth.
2. The network domain: The network domain consists of the information reflecting the network status such as traffic, system load, network revenue, network delay, routing, scheduling scheme and node topology.
3. The user domain: This domain focuses on information of concern to users, such as location information, quality of service (QoS) preference, user identities (IDs) and accounting.
4. The policy domain: Policy is the guideline that manages radio resources such as communication rules and spectrum policy.

The logical framework of a multi-domain cognition database is shown in Fig. 4.6, which includes a local database and a cooperative database. The local database can store and manage multi-domain information from the CWN in certain regions. Local databases deployed in various regions can provide the multi-domain information interaction via cooperations. The learning ability in the CWN can interpret the multi-domain information into knowledge and cases, leading to a smarter and more credible intelligent prediction.

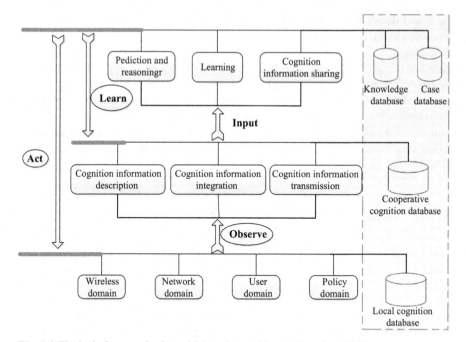

Fig. 4.6 The logic framework of a multi-domain cognition database in a CWN

4.2.3   Three Region Schemes with Combined Spectrum Sensing

The exploration and the exploitation of available spectrum opportunities are to increase the opportunity of spectrum access for secondary users (SUs), namely, to improve the spectrum utilization and the capacity of SUs. To utilize the spectrum holes, the first step is to find them in the space and time domains. Therefore, a novel three region scheme for space-time spectrum sensing and access is proposed, which includes the black region, grey region and white region. And the bounds of three regions are calculated by taking into account key interference factors from SUs and the miss detection and false alarm probabilities in spectrum sensing that aims to discover space-time spectrum holes. As illustrated in Fig. 4.7, the network model with PUs and SUs located in a disc of radius R is considered. A primary transmitter, for example, a TV station is located in the center of a circle and transmits with power P_0. Primary users, such as TV receivers, are distributed uniformly around the primary transmitter within the black region. The whole disc is divided into three regions, i.e. black region with outer radius R_1, grey region with outer radius R_2, white region with inner radius R_3 and outer radius R, where R could be infinite. The SUs are distributed uniformly in the grey and white region with density λ users per unit area and transmit with power P_s. The shadow area is the transition zone, where SUs need to control their transmit power to mitigate the interference to primary receivers.

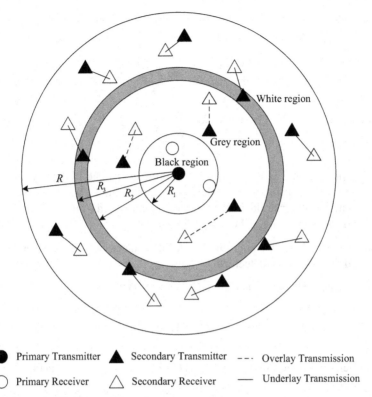

Fig. 4.7 System model of three region scheme

1) Temporal spectrum opportunity in grey region

In grey region, since SUs operate in overlay spectrum sharing, they are required to sense the radio spectrum environment and detect the spectrum bands which are not occupied by the PUs. Here, energy detection scheme is used to sense the spectrum. Then, the probabilities of miss detection $P_m = \Pr\{H_0|H_1\}$ and false alarm $P_f = \Pr\{H_0|H_1\}$ are denoted in Eq. (4.14).

$$p_m(\varepsilon,\tau) = 1 - p_d(\varepsilon,\tau) = Q\left(\left(\gamma+1-\frac{\varepsilon}{\sigma_w^2}\right)\sqrt{\frac{N}{2\gamma+1}}\right)$$

$$p_f(\varepsilon,\tau) = Q\left(\left(\frac{\varepsilon}{\sigma_w^2}-1\right)\sqrt{\tau f_s}\right) = Q\left(\left(\frac{\varepsilon}{\sigma_w^2}-1\right)\sqrt{N}\right) \quad (4.14)$$

N is the number of samples, γ is the signal to noise ratio (SNR) at the detector, σ_w^2 is the power spectral density of AWGN and ε is the decision threshold. P_m and P_f are denoted in Eq. (4.15).

$$p_m(\varepsilon,N) \le \xi_m$$

$$p_f(\varepsilon,N) \le \xi_f \quad (4.15)$$

Substituting (4.16) into (4.15), solving the equations yields two constraints of the detection threshold ε.

$$\varepsilon \leq \sigma_w^2 \left(\gamma + 1 - Q^{-1}(\xi_m) \sqrt{\frac{2\gamma+1}{N}} \right) \triangleq \varepsilon_m$$

$$\varepsilon \geq \sigma_w^2 \left(\frac{Q^{-1}(\xi_f)}{\sqrt{N}} + 1 \right) \triangleq \varepsilon_f \tag{4.16}$$

$Q^{-1}(\cdot)$ is the inverse Q function. Two key parameters ε_m and ε_f determine the range of ε to satisfy the constraint of P_m, the detection threshold ε must be lower than ε_m while to satisfy the constraint of P_p, the detection threshold ε must be higher than ε_f. Therefore, the detection threshold is $\varepsilon \in \left[\varepsilon_f, \varepsilon_m \right]$.

Theorem 1 ε_m is a decreasing function of the distance between the primary transmitter and the secondary detector.

Proof The derivative of ε_m is denoted in Eq. (4.17).

$$\frac{d\varepsilon_m}{dr} = \sigma_w^2 \left(1 - Q^{-1}(\xi_m) \sqrt{\frac{2}{N}} \right) \tag{4.17}$$

For sufficiently large N, when $N > 2 \left(Q^{-1}(\xi_m) \right)^2$, results are achieved as denoted by $Q^{-1}(\xi_m) \sqrt{\frac{2}{N}} < 1$ and $\frac{d\varepsilon_m}{dr} > 0$. For example, let $\xi_m = 10^{-3}$, then when $N \geq 20, \frac{d\varepsilon_m}{dr} > 0$. Actually, Eq. (4.14) is derived by the Central Limit Theorem (CLT), and to satisfy CLT, N must be sufficiently large.

Since SNR $\gamma = \frac{P_0/L^\alpha}{\sigma_w^2}$ is a decreasing function of L, where L is the distance between primary transmitter and secondary detector, $\frac{d\varepsilon_m}{dL} = \frac{d\varepsilon_m}{dr} \frac{dr}{dL} < 0$. And this theorem is proved.

The SU takes the decision threshold ε from the interval $\varepsilon \in \left[\varepsilon_f, \varepsilon_m \right]$, where ε_f is a constant and ε_m is a decreasing function of the distance r. As ε_m decreases, there must be an r, where $\varepsilon_f < \varepsilon_m$ and the interval $\left[\varepsilon_f, \varepsilon_m \right]$ is empty. Thus the secondary user cannot find a detection threshold for spectrum sensing. Therefore $\varepsilon_f = \varepsilon_m$ reveals the limit of temporal spectrum sensing. Solving this equation yields the bound of grey region R_2.

$$R_2 = \left(\frac{P_0 N}{\sigma_w^2 \left(q_f \sqrt{N} + q_m^2 + q_m \sqrt{N + 2q_f \sqrt{N} + q_m^2} \right)} \right)^{1/\alpha} \tag{4.18}$$

Fig. 4.8 Worst interference from SUs to PU

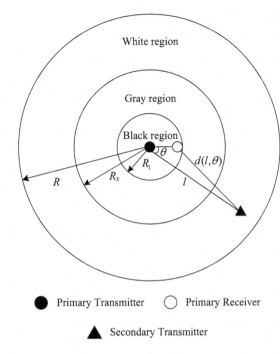

Primary Transmitter ●   ○ Primary Receiver

▲ Secondary Transmitter

2) Spatial spectrum opportunity in white region

i) Accurate method: The PUs on the edge of black region suffer from the worst inter-ference generated by SUs. Assume all SUs transmit with power P_s. The interference experienced by PUs on the edge of black region from the SUs in white region is depicted in Eq. (4.19).

$$I(r,\theta) = \frac{P_s}{[d(r,\theta)]^\alpha} = \frac{P_s}{\left(r^2 + R_1^2 - 2R_1 r \cos\theta\right)^{\alpha/2}} \qquad (4.19)$$

$d(r,\theta)$ is the distance from SU to PU on the edge of black region. r and θ are depicted in Fig. 4.8. As all SUs are distributed uniformly, both r and θ are random variables (r.v.) with probability density functions (PDF) in Eqs. (4.20) and (4.21).

$$f_r(r) = \frac{2r}{R^2 - R_x^2}, R_x < r \le R \qquad (4.20)$$

$$f_\theta(\theta) = \frac{1}{2\pi}, 0 < \theta \le 2\pi \qquad (4.21)$$

Denote I_0 as the aggregated interference inflicted on PU by all SUs, and the expectation is denoted in Eq. (4.22).

$$E[I_0] = \int_{R_x}^{R} \int_0^{2\pi} \frac{\lambda r P_s}{\left(r^2 + R_1^2 - 2R_1 r \cos\theta \right)^{\alpha/2}} \, d\theta dr \tag{4.22}$$

When $\alpha = 2k$, with k as a positive integer, a closed-form of interference can be derived as Eq. (4.23).

$$E[I_0] = \int_{R_x}^{R} \frac{\lambda r P_s}{R_1^{2k}} \int_0^{2\pi} \frac{1}{\left(1 - 2a\cos\theta + a^2 \right)^k} \, d\theta dr$$

$$= \frac{\lambda P_s}{R_1^{2k-2}} \int_{R_x/R_1}^{R/R_1} a \int_0^{2\pi} \frac{1}{\left(1 - 2a\cos\theta + a^2 \right)^k} \, d\theta da \tag{4.23}$$

ii) Supplement method: By recentering the network at the primary receiver on the edge of black region and fill the annulus whose radii range from $R_x - R_1$ to $R_1 + R$ with secondary transmitters with the same density λ users per unit area, it is also assumed that $R_x - R_1 > 1$ and the singular point can be removed from the integral. The upper bound of interference and its limit are shown in Eqs. (4.24) and (4.25), when $R \to \infty$.

$$E[I_0] \le 2\pi P_s \lambda \int_{R_x - R_1}^{R+R_1} r^{1-\alpha} dr = \frac{2\pi P_s \lambda}{\alpha - 2} \left(\frac{1}{(R_x - R_1)^{\alpha-2}} - \frac{1}{(R+R_1)^{\alpha-2}} \right) \tag{4.24}$$

$$\lim_{R \to \infty} E[I_0] \le \frac{2\pi P_s \lambda}{\alpha - 2} \frac{1}{(R_x - R_1)^{\alpha-2}} \tag{4.25}$$

3)   The bounds of three regions

The outage constraints of PU are adopted to constraint the radius of black region R_1. Assume the data rate of a PU is T_0, and then the outage event occurs when T_0 falls below a threshold data rate C_0. To guarantee the QoS of PU, the outage probability cannot exceed β, which is the worst interference in Eq. (4.26).

$$P_{out} = \Pr\left[\log\left(1 + \frac{P_0/R_1^\alpha}{I_0 + \sigma_w^2} \right) \le C_0 \right] \le \beta = \Pr\left[I_0 \ge \frac{P_0/R_1^\alpha}{2^{C_0} - 1} - \sigma_w^2 \right] \le \beta \tag{4.26}$$

Define $I_{th} = \frac{P_0/R_1^\alpha}{2^{C_0} - 1} - \sigma_w^2$, which is an interference threshold. When $I_0 \ge I_{th}$, an outage of PU occurs. It is noted by $I_{th} \ge 0$.

$$R_1 \leq \left(\frac{P_0}{\sigma^2(2^{C_0}-1)}\right)^{1/\alpha} = R_1^u \tag{4.27}$$

R_1^u is the upper bound of the radius of black region. Thus, we have $R_1 \leq R_1^u$.

The radius of white region can be derived by the outage constraint of PU as depicted in Eq. (4.28).

$$P_{out} \leq \frac{E[I_0]}{I_{th}} = \frac{E[I_0]}{\dfrac{P_0/R_1^\alpha}{2^{C_0}-1}-\sigma^2} \tag{4.28}$$

When $\alpha = 4, R \to \infty$, using β to bound the right side of the inequality in Eq. (4.29).

$$R_x \geq \left(\left(\frac{\pi\lambda P_s R_1^2}{\beta\left(\dfrac{P_0/R_1^\alpha}{2^{C_0}-1}-\sigma^2\right)}\right)^{1/2} + R_1^2\right)^{1/2} \overset{def}{=} R_c \tag{4.29}$$

It is noticed that all R_c above are increasing functions of P_s, because when P_s increases, SUs should be farther away from PUs to reduce the interference and R_c is thus increased. The radius of white region is $R_3 = \max\{R_c, R_2\}$. The radii of three regions are summarized as follows.

- The radius of black region is $R_1 < R_u$
- The radius of grey region is R_2 as Eq. (4.18).
- The radius of white region is $R_3 = \max\{R_c, R_2\}$.

4)   Numerical results and analysis

The existence of a transition zone is illustrated in Fig. 4.9, where a family of curves with different values of C are presented. When the pair $(\varepsilon_f; \varepsilon_m)$ is below the curve, the transition zone exists. Take $(\varepsilon_f; \varepsilon_m) = (0:2; 0:1)$ as an example, when $C=4:5$, the transition zone is eliminated, but when $C=4$, the transition zone exists. Figure 4.9 shows that the relaxation of ε_f or ε_m can eliminate the transition zone.

Figure 4.10 shows the relation between R_3 and R_1 with P_s as the reference variable. The larger the R_1, the larger the R_3, for the SUs in white region need to make a concession when the black region is expanded. Besides, R_3 is a convex function of R_1, which depicts that R_3 increases faster when R_1 is larger.

Figure 4.11 displays the relation between R_1 and R_3 with P_0 as the reference variable. The larger the P_0, the smaller the R_3, due to the fact that the signal strength of PU increases. And the SNR of PU improves as the white region expands inward. Notice that R_3 is also a convex function of R_1. When R_1 is small, $R_3 = R_2$, which denotes that the transition zone is eliminated.

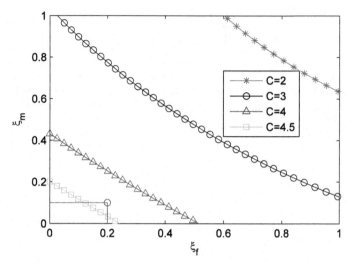

Fig. 4.9 The existence of a transition zone

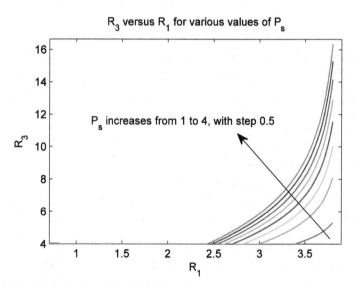

Fig. 4.10 The relation between $R1$ and $R3$ for various values of Ps

The relation between P_s and the width of the transition zone is illustrated in Fig. 4.12. The larger the P_s, the wider the transition zone, because the interference from SUs is larger than before and the transition zone needs to expand to protect primary receivers. Notice that the larger P_0, the narrower the transition zone, be-

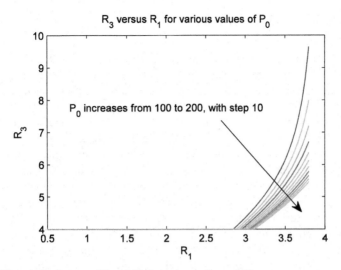

Fig. 4.11 The relation between $R1$ and $R3$ for various values of $P0$

cause the SNR of primary receiver improves and primary receiver can tolerate more interference, leading to the reduction of the radius of white region and the shrink of the transition zone. Notice that by tuning parameters P_s and P_0, the transition zone can also be eliminated, i.e. when P_s is sufficiently small or P_0 is large enough, the transition zone will disappear.

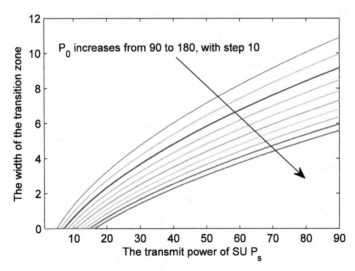

Fig. 4.12 The relation between P_s and the width of the transition zone for various values of P_0

4.3   Cognitive Pilot Channel

As analyzed above, the cognitive ability in CWNs requires the acquisition of network information. Spectrum sensing provides a way to collect and store the spectrum and network information. However, there is still a big challenge on how to design the signaling channel to transmit the network cognition information accurately and efficiently among heterogeneous networks in CWNs. Therefore, the concept of cognitive pilot channel (CPC), proposed in the E2R and E3 projects [33], is considered as one of the candidate signaling channel technologies in CWNs. The challenges of CPC technology are summarized in two main aspects: one is the architecture of the CPC deployment including operation mode (in-band mode or out-band mode), information delivery mode (broadcast mode or on-demand mode) and information structuring scheme (single layer or multiple layer). Other aspects for the efficient cognitive information delivery using CPC technologies are described in [34–38]. Recently, 3GPP standard has paid much attention on the control signaling design by proposing the dual connectivity technology for data traffic offloading in the small cell enhancement in LTE release 12 in [39].

4.3.1   Geographically Homogeneous Mesh Grouping Scheme for Broadcast CPC

Since the information of CPC is categorized by meshes, the mesh management scheme is important in CPC deployment which has not been considered yet. In [40], mesh division problem is proposed and the author suggests mesh with an adaptive size. The research on the optimal mesh division scheme is proposed in [36] and [37]. The effects of the GPS localization shift error and information loss ratio in the multi-RATs (multiple Radio Access Technologies) overlapped scenario is considered in [36] which can improve the accuracy and efficiency of network information delivery. A novel MAC layer frame for mesh grouping is proposed by [37]. However, the meshes are not geographically grouped in these literatures. In this section, an adaptive size mesh grouping scheme is proposed, which can adapt to the radio environment and significantly reduce the number of meshes.

1. Homogeneous mesh grouping algorithm
A mesh is defined as an area with the same subset of RATs. The goal of mesh grouping algorithm is to group the geographic area with the same radio parameters into one mesh. In this paper, the shape of a mesh is rectangle. Thus, the algorithm is to group the area with the same subset of RATs into one rectangle.

Two algorithms are designed here. The first one is to divide the unqualified mesh and retain the qualified mesh iteratively, where "qualified mesh" is the mesh whose radio environment is almost pure. Because this algorithm adopts the iteration operation in the fractal theory, it is denoted as FbMDA (Fractal Based Mesh Division

Fig. 4.13 Mesh division
demonstration

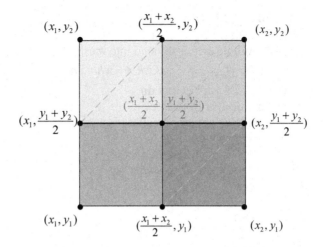

Algorithm). After mesh fusion algorithm (denoted as MFA-aFbMD, Mesh Fusion Algorithm after Fractal based Mesh Division), the homogeneous meshes are further fused. The first algorithm is denoted as FbMD-MGA (Fractal based Mesh Division and Mesh Grouping Algorithm), which is FbMDA followed by MFA-aFbMD.

The second algorithm aims at dividing the geographic area into regular meshes (denoted as RMD, Regular Mesh Division), then fuses the regular meshes into larger irregular meshes (denoted as MFA-aRMD, Mesh Fusion Algorithm after Regular Mesh Division). The second mesh grouping algorithm is denoted as RMD-MGA (Regular Mesh Division and Mesh Grouping Algorithm), which is RMD followed by MFA-aRMD. Mesh grouping algorithm is always mesh division followed by mesh fusion operation.

2. FbMDA and MFA-aRMD algorithm

FbMDA Description FbMDA is described in Algorithm 1, where each mesh is denoted by the coordinates of vertexes on the diagonal, as well as the whole area, which is an unqualified mesh. The division operation is illustrated in Fig. 4.13, mesh $[x_1, y_1; x_2, y_2]$ is divided into four smaller meshes as

$$\left[x_1, \frac{y_1+y_2}{2}; \frac{x_1+x_2}{2}, y_2\right], \left[\frac{x_1+x_2}{2}, \frac{y_1+y_2}{2}; x_2, y_2\right],$$
$$\left[x_1, y_1; \frac{x_1+x_2}{2}, \frac{y_1+y_2}{2}\right], \left[\frac{x_1+x_2}{2}, y_1; x_2, \frac{y_1+y_2}{2}\right]$$

Homogeneous meshes after FbMDA do exist. So FbMDA followed by MFA (Mesh Fusion Algorithm) brings better performance. FbMD (MFA-aFbMD) is described in Algorithm 2, which is different from MFA after regular mesh division, as the data structures of two algorithms are different.

Algorithm 1 FbMDA

Require: Discrete processing is finished.
Ensure: Whole area is divided into meshes.
1: Initialize σ_{max} , e_{min} , *stack* $\leftarrow$ *whole area, mesh* $\leftarrow \varnothing$
2: **While** *stack* isn't empty **do**
3:　　Pop mesh *m* from *stack*
4:　　Divide *m*, get four smaller meshes $m_1, \ldots, m_4$.
5:　　**for** *j*=1 to 4 **do**
6:　　　**if** m_j is qualified **then**
7:　　　　Append m_j to *mesh*
8:　　　**else**
9:　　　　Push m_j in *stack*
10:　　　**end if**
11:　　**end for**
12: **end while**

'

Require: FbMD is finished.
Ensure: Homogeneous meshes after FbMD are fused.
1: Get mesh number *L, label* $\leftarrow$ *true*
2: **while** *label* **do**
3:　*label* $\leftarrow$ *false*, $i \leftarrow 1$, get $mesh_i$
4:　**while** $i \leq L$ **do**
5:　　$j \leftarrow 1$
6:　　**while** $j \leq L$ **do**
7:　　**if** $mesh_j \leftrightarrow mesh_i$ **then**
8:　　　Fuse $mesh_i$ and $mesh_j$
9:　　　Update array *mesh, L=L*-1, *label* $\leftarrow$ *true*
10:　　**end if**
11:　　$j=j+1$
12:　　**end while**
13:　　$i=i+1$
14:　**end while**
15: **end while**

If two meshes are fusible, they are not only geographically adjacent, but also have the same radio environment. Fusing mesh *i* and mesh *j* into one is an operation that fuses two rectangles with the same edge into one rectangle. As illustrated in Fig. 4.14, mesh $[x_1, y_1; x_2, y_2]$ and $[x_3, y_3; x_4, y_4]$ are fused into mesh $\left[\min_{i=1\ldots4} x_i, \min_{i=1\ldots4} y_i ; \max_{i=1\ldots4} x_i, \max_{i=1\ldots4} y_i \right]$. If mesh *i* and mesh *j* are fusible, the relation is denoted as "mesh $i \leftrightarrow$ mesh j".

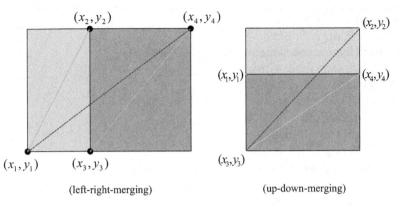

(left-right-merging) (up-down-merging)

Fig. 4.14 Mesh fusion demonstration

RMD-MGA Description After regular mesh division, meshes are stored in a regular order. Thus, more efficient data structure can be designed to save the searching time. As illustrated in Fig. 4.17, array map stores the right-adjacent and under-adjacent neighbors of a mesh, where the element 0 denotes an invalid mesh. Take ith column as an example, map $[1, i]$ is the index of mesh i, map$[2, i]$ is the index of right-adjacent mesh for mesh i, and map $[3, i]$ is the index of under-adjacent mesh for mesh i. As the adjacent mesh is stored in this ordered way, searching time is reduced significantly. Some algorithm parameters are defined here: the whole area is divided into $N_m = N_r \times N_c$ meshes, where N_r, N_c are the number of rows and columns respectively. The feature of each mesh is stored in an array $[f]_{N_m \times 1}$. MFA-aRMD is described in

Algorithm 3, whose key operation is the update of mesh's neighbors after fusion. The algorithm is denoted as RMD-MGA (Regular Mesh Division and Mesh Grouping Algorithm), which is RMD followed by MFA-aRMD (Fig. 4.15).

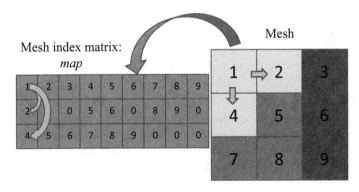

Fig. 4.15 Data structure of MFA-aRMD

Algorithm 3 MFA-aRMD

Require: RMD is finished.

Ensure: Homogeneous meshes after RMD are fused.

1: Initialize *map*, *label* ← *true*,

2: **While** *label* **do**

3:   *label* ← *false*

4: **for** *i*=1 to N_m **do**

5:   **if** *map* [1, *i*] ≠ 0 **then**

6:     L_r ← *map* [2, *i*]

7:     **if** L_r ≠ 0 & $mesh_i$ ↔ $mesh_{L_r}$ **then**

8:       *map* [2, *i*] ← *map* [2, L_r], *map* [1, L_r] ← 0

9:       **for** *j*=1 to L_r -1 **do**

10:         **if** *map* [1, *j*] ≠ 0 & *map* [3, *j*] = L_r **then**

11:           *map* [3, *j*] ← *i*

12:         **end if**

13:       **end for**

14:     Update array *mesh*, *label* ← *true*

15:     **end if**

16:     L_d ← *map* [3, *i*]

17:     **if** L_d ≠ 0 & $mesh_i$ ↔ $mesh_{Ld}$ **then**

18:       *map* [3, *i*] ← *map* [3, L_d], *map* [1, L_d] ← 0

19:       **for** *j*=1 to L_d -1 **do**

20:         **if** *map* [1, *j*] ≠ 0 & *map* [2, *j*] = L_d **then**

21:           *map* [2, *j*] ← *i*

22:         **end if**

23:       **end for**

24:     Update array *mesh*, *label* ← *true*

25:     **end if**

26:   **end if**

27: **end for**

28: **end while**

---

3. Simulation results

First, evaluation parameters for these two algorithms are proposed. The Average Radio Parameters Error is denoted as ARPE. The Radio Parameters Error of mesh *i* is denoted as RPEi, then two definitions of ARPE are denoted in Eq. (4.30), which can evaluate the performance of mesh grouping algorithms.

$$ARPE_1 = \frac{1}{S_a} \sum_{i=1}^{N_m} S_i \times RPE$$

$$ARPE_2 = \frac{1}{N_m} \sum_{i=1}^{N_m} RPE_i \tag{4.30}$$

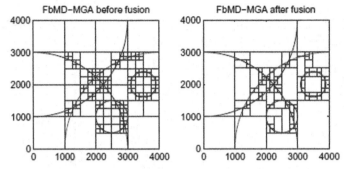

Fig. 4.16 Results of FbMD-MGA

S_a is the area of the whole geographic region, S_i is the area of mesh i and N_m is the number of meshes.

Let $N = 1024$, $\sigma_{max} = 10\%$, $e_{min} = 16$ pixel , where e_{pixel} is the length of a pixel's edge. Set the time to transmit the information of a mesh as 20 ms, and user density range from 11 to 2011 users per km². The simulation results are illustrated in Figs. 4.16, 4.17, 4.18, and 4.19.

Figure 4.16 is the mesh division result of FbMDA. After the mesh fusion, the number of meshes is reduced significantly. Figure 4.17 is the result of RMD, and the number of meshes is large. After the mesh fusion, the number of meshes is also reduced. Figure 4.18 gives the curve of ARPE versus mesh number and its linear fit of RMD-MGA's and FbMD-MGA's respectively. It's obvious that ARPE reduces as mesh number increases. And the performance of two algorithms is evaluated by different curves. When the mesh number is 50, RMD-MGA's $ARPE_1$ and $ARPE_2$ are both smaller than that of FbMD-MGA's. Besides, the slope of the linear fit of $ARPE_1$ in RMD-MGA is greater than that in FbMD-MGA, which depicts that $ARPE_1$ of RMDMGA reduces faster than that of FbMD-MGA when mesh number increases. In this means, RMD-MGA's performance is better than FbMD-MGA. But note that RMD-MGA requires regular mesh division, when the whole area is irregular, the performance of FbMD-MGA is conversely better.

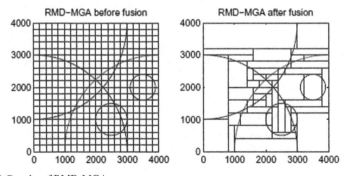

Fig. 4.17 Results of RMD-MGA

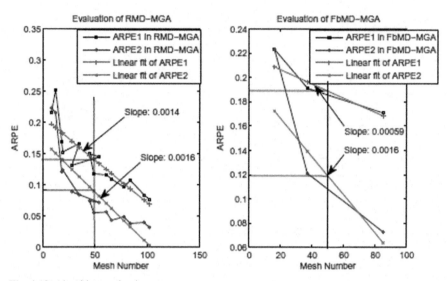

Fig. 4.18 Algorithm evaluation

The average delay of each user in receiving the broadcast information is illustrated in Fig. 4.19, which denotes that fusion operation brings benefits to the delay reduction. Besides, as the number of meshes in RMD-MGA is smaller than that in FbM-MGA, the delay of RMD-MGA is also smaller than that of FbMD-MGA.

In conclusion, the novel mesh grouping scheme can significantly reduce the number of meshes by using the meshes grouping scheme. Meanwhile, the average

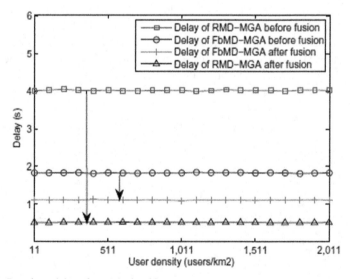

Fig. 4.19 Broadcast delay after each algorithm

delay in receiving CPC information is also reduced. Mesh grouping scheme provides an efficient data fusion mechanism for cognitive database.

4.3.2   Efficient Mesh Division and Differential Information Coding Schemes in Broadcast CPC

As one of the delivery mode of CPC, the original broadcast CPC mode is proposed as a narrow-band broadcast channel for limited network information delivery with a large coverage. Further design of the broadcast CPC mode is presented which will broadcast the network information of all meshes periodically and continuously. The flow of original broadcast CPC mode is shown in Fig. 4.20, which depicts that the network information of all meshes is broadcast one by one periodically. Within each mesh, different fields are encoded with different numbers of bits to represent the network information, such as the Geographical field (Geo field for short), Operator field, RAT field and Frequency field.

The advantages of broadcast CPC mode include the large coverage, only downlink channel and easy implementation for UEs which would just switch on and listen to the broadcast information. However, potential problems still exist when the information corresponding to a specific mesh is missed by the UE, which has to wait for a whole broadcast period to catch the information again. The network information redundancy of adjacent meshes will consume a large portion of specious spectrum resources and will cost the unnecessary time delay in the broadcast CPC mode. These are the pros and cons of the original broadcast CPC mode.

In contrast to the original broadcast CPC mode which broadcast the network information of each mesh totally, the proposed differential information coding (DIC) scheme will choose a basic mesh with popular commonality as a reference and quantize the differential information of other meshes against the basic mesh. Other meshes will make a reference to the basic mesh and only the differential information will be encoded and transmitted through the narrowband broadcast CPC channel, greatly improving the efficiency of information delivery.

The original broadcast CPC mode coding scheme is shown in Fig. 4.21 which includes all the network information in each mesh. The network information in Mesh #i, #j and #k as M_i, M_j, and M_k, is shown as an example to interpret the difference

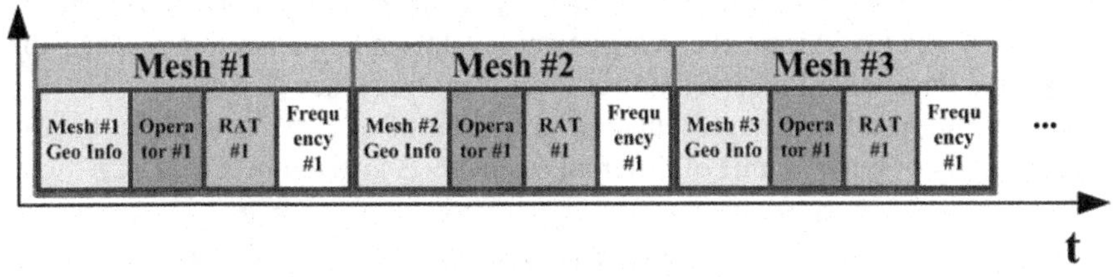

Fig. 4.20 Flow of the original broadcast CPC Mode

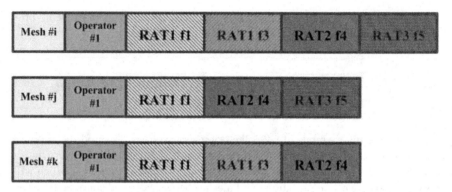

Fig. 4.21 Original broadcast CPC mode coding

between original and DIC schemes. The proposed DIC scheme using differential coding strategy is shown in Fig. 4.22, which only transmit the differential information in each mesh against the basic mesh, saving spectrum resources and improving the broadcast efficiency. To be specific, M_i is chosen as the basic mesh and M_j only transmits the differential information of deleting RAT1 with f_3 comparing to the whole original information list such as the RAT1 with f_1, RAT2 with f_4 and RAT3 with f_5. The same scheme applies to M_k which only transmits the differential information of deleting RAT3 with f_5 comparing to the network information in the basic mesh. By using the DIC scheme, the most commonly appeared network information is transmitted in the basic mesh and other meshes only transmit the differential information comparing to the basic mesh, which greatly saving spectrum resources and improving the information coding efficiency in the broadcast CPC mode.

Based on the image processing techniques and analysis, the frequency occupancy condition is first digitalized with binary values and then depicted in different colors to represent the frequency occupancy condition. It is also assumed that the received power levels of each RAT in each point are known in the first place. If the received power level of frequency f_i in point (x, y) is above the threshold, the frequency occupancy graph for this point will be marked with binary 1 to represent the coverage existence of frequency f_i. The formula is shown in Eq. (4.31).

$$M(f_i, x, y) = \begin{cases} 1, & \text{if } P(f_i, x, y) \geq P_{th}(f_i) \\ 0, & \text{if } P(f_i, x, y) < P_{th}(f_i) \end{cases} \tag{4.31}$$

where $M(f_i, x, y)$ is the binary value of frequency occupancy condition of frequency f_i in point (x, y). $P(f_i, x, y)$ is the receiving power level from frequency f_i in point (x, y). $P_{th}(f_i)$ is the threshold of received power level from frequency f_i.

By marking the frequency occupancy condition of each point with the binary value, the sum of binary value of different frequencies is given by Eq. (4.32).

$$I(x, y) = \sum_{i=1}^{N} \left[M(f_i, x, y) 2^{i-1} \right] \tag{4.32}$$

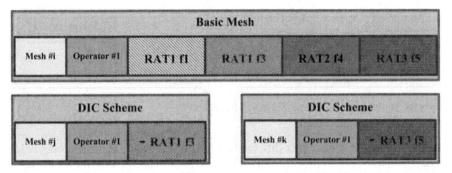

Fig. 4.22 DIC scheme

where $I(x, y)$ depicts the sum of frequency occupancy in point (x, y). N is the total number of frequencies. And the total number of levels for I is depicted by $L_I = 2^N$. For example, when $N = 5$ frequencies f_1, f_2, f_3, f_4, f_5, a point covered by 5 frequencies will be marked with $I = (11111)_2 = (31)_{10}$.

Let us define the digitalized value for each mesh. Parameter I_{Mk} is used to depict the value of different frequencies in M_k. Parameter N_{Mk} depicts the total number of points in M_k. In order to avoid the inaccuracy of the scheme by using the average value I of all points in each mesh, this chapter proposes a much more accurate approach by choosing the value I with the greatest proportion in each mesh to represent the frequency occupancy status of each mesh. The range of I is depicted by $I \in \{0, 1, \ldots, (L_I - 1)\}$. The proportion of value $I = j$ in M_k is depicted by $P_{M_k}^j$ in Eq. (4.33), where the N_{Ij} depicts the number of points with the value $I = j$. The mesh value for M_k is depicted by I_{Mk} in Eqs. (4.33) and (4.34) with the greatest proportion $P_{M_k}^j$.

$$P_{M_k}^j = \frac{N_{Ij}}{N_{M_k}} \tag{4.33}$$

$$I_{M_k} = \{j \mid j = \max_{j \in \{0,1,\ldots(L_I - 1)\}} P_{M_k}^j\} \tag{4.34}$$

Based on the DIC scheme, further improvements of the frame format are proposed, including the basic mesh frame format and the differential mesh frame format in Figs. 4.23 and 4.24. The Basic Mesh Flag field is used to distinguish the basic mesh from others with $B_{BMF} = 1$bit, where '1' depicts the basic mesh frame format

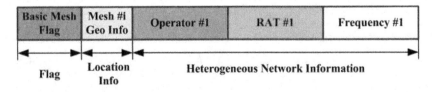

Fig. 4.23 Basic mesh frame format of DIC scheme

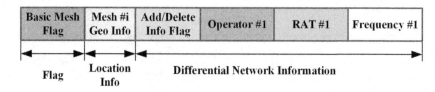

Fig. 4.24 Differential mesh frame format of DIC scheme

in Fig. 4.23 and '0' depicts the differential meshes. The Add/Delete Info Flag field is designed to represent the differential network information against the basic mesh with $B_{ADF} = 1$ bit, where '1' for adding differential information and '0' for deleting differential information in contrast to the basic mesh. All the UEs in each mesh will first decode the basic mesh information and then add or delete the differential network information based on its location.

In order to calculate the total number of bits to encode the network information with different operators, RATs and frequencies in M_j, general formulas are represented by $B_{M_j}^{ORG}$ for the original broadcast mode in Eq. (4.35), $B_{M_j}^{BM}$ for the basic mesh in Eq. (4.36) and $B_{M_j}^{DM}$ for the differential meshes in Eq. (4.37). Suppose the number of operators in M_j is depicted by $N_{OP,j}$ with $N_{RAT,j}^{OP,k}$ RATs of the kth operator and $N_{FREQ,j}^{RAT,t}$ of the tth RAT. Furthermore, the total number of bits for all meshes are depicted by B_{Total}^{ORG} for the original broadcast mode in Eq. (4.38) and B_{Total}^{DIC} for the DIC scheme in Eq. (4.39), where N_{MESH} is the total number of meshes.

$$B_{M_j}^{ORG} = B_{GEO} + \sum_{k=1}^{N_{OP,j}} \left(B_{OP} + \sum_{t=1}^{N_{RAT,j}^{OP,k}} \left(B_{RAT} + \sum_{s=1}^{N_{FREQ,j}^{RAT,t}} B_{FREQ} \right) \right) \qquad (4.35)$$

$$B_{M_j}^{BM} = B_{BMF} + B_{GEO} + \sum_{k=1}^{N_{OP,j}} \left(B_{OP} + \sum_{t=1}^{N_{RAT,j}^{OP,k}} \left(B_{RAT} + \sum_{s=1}^{N_{FREQ,j}^{RAT,t}} B_{FREQ} \right) \right) \qquad (4.36)$$

$$B_{M_j}^{DM} = B_{BMF} + B_{GEO} + \sum_{k=1}^{N_{OP,j}} \left(B_{ADD} + B_{OP} + \sum_{t=1}^{N_{RAT,j}^{OP,k}} \left(B_{RAT} + \sum_{s=1}^{N_{FREQ,j}^{RAT,t}} B_{FREQ} \right) \right) +$$

$$\sum_{k=1}^{N_{OP,j}} \left(B_{DEL} + B_{OP} + \sum_{t=1}^{N_{RAT,j}^{OP,k}} \left(B_{RAT} + \sum_{s=1}^{N_{FREQ,j}^{RAT,t}} B_{FREQ} \right) \right) \qquad (4.37)$$

$$B_{Total}^{ORG} = \sum_{j=1}^{N_{MESH}} B_{M_j}^{ORG} \qquad (4.38)$$

$$B_{Total}^{DIC} = \sum_{j=1}^{N_{MESH}} B_{M_j}^{BM} + \sum_{k=1}^{N_{MESH}} B_{M_k}^{DM} \qquad (4.39)$$

Next, the Homogeneous Meshes Grouping (HoMG) based DIC scheme is proposed to further improve the efficiency of broadcast CPC mode by reducing the duplicate network information iteration in the original broadcast CPC mode, especially for the scenario that the network information of adjacent meshes are the same and unnecessarily delivered repetitively.

The meshes with the same digitalized value $I_M k$ in Eq. (4.34) are detected and filtered out by the appropriate threshold.

$$N_G = 2^N \qquad (4.40)$$

$$G_t \underset{t \in \{0, N_G-1\}}{=} \left\{ M_i \middle| I_{M_i} = t \right\} \qquad (4.41)$$

To be specific, the total number of groups is represented by N_G in Eq. (4.40), where N denotes the total number of frequencies. The grouping set of homogeneous meshes is represented by G, where t can be set from 0 to (N_G-1) depicting the digitalized value of the group and only the meshes with the same value t can be marked as the members of G_t in (4.41). And parameter $i \left(1 \le i \le N_{MESH}\right)$ is the index of all meshes.

Furthermore, based on the analyses and formulas for DIC scheme, the HoMG scheme is brought forward to further decrease the redundancy of homogeneous meshes information repetition. Based on the DIC scheme in Eqs. (4.35) and (4.39), the HoMG based DIC scheme formulas are shown in Eqs. (4.42), (4.43) and (4.44). The basic mesh group coding scheme for the jth group is shown in Eq. (4.42) by $B_{G_j}^{BM}$, where $N_{G_j} \ne 0$. And the differential mesh group coding scheme for the jth group is shown in Eqs. (4.43) by $B_{G_j}^{DM}$, where $N_{G_j} \ne 0$. If $N_{G_j} = 0$, then there is no need to do the differential mesh group coding for the jth group. Finally, the total number of bits for the HoMG based DIC scheme is shown in Eqs. (4.44) by $B_{Total}^{HoMG\_DIC}$.

$$B_{G_j}^{BM} = B_{BMF} + B_{GEO} N_{G_j} + \sum_{k=1}^{N_{OP,j}} \left(B_{OP} + \sum_{t=1}^{N_{RAT,j}^{OP,k}} \left(B_{RAT} + \sum_{s=1}^{N_{FREQ,j}^{RAT,t}} B_{FREQ}\right)\right) \qquad (4.42)$$

$$B_{G_j}^{DM} = B_{BMF} + B_{GEO} N_{G_j} + \sum_{k=1}^{N_{OP,j}} \left(B_{ADD} + B_{OP} + \sum_{t=1}^{N_{RAT,j}^{OP,k}} \left(B_{RAT} + \sum_{s=1}^{N_{FREQ,j}^{RAT,t}} B_{FREQ}\right)\right) +$$

$$\sum_{k=1}^{N_{OP,j}} \left(B_{DEL} + B_{OP} + \sum_{t=1}^{N_{RAT,j}^{OP,k}} \left(B_{RAT} + \sum_{s=1}^{N_{FREQ,j}^{RAT,t}} B_{FREQ}\right)\right) \qquad (4.43)$$

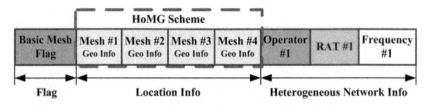

Fig. 4.25 Basic mesh frame format of HoMG based DIC scheme

$$B_{Total}^{HoMG\_DIC} = \sum_{k=1}^{N_G}(B_{G_j}^{BM} + B_{G_j}^{DM}) \tag{4.44}$$

The frame formats for both the basic mesh and differential mesh of HoMG based DIC schemes are designed in Figs. 4.25 and 4.26 in contrast to the DIC scheme in Figs. 4.23 and 4.24. The major differences of the HoMG based DIC scheme are that it groups homogeneous meshes together with Geo Info fields and sends the network information only once, reducing the unnecessary duplicate information delivery in the original mode.

Then, based on the formulas in Eqs. (4.42), (4.43) and (4.44), the HoMG based DIC scheme further improves the efficiency of network information coding by grouping the homogeneous meshes together. As shown in Fig. 4.27, the total number of bits for encoding the network information is compared between the DIC scheme and the HoMG based DIC scheme. Performance improvements of the data coding compression are achieved by decreasing the duplicate network information delivery in Fig. 4.27. Specifically, a proportion of 60 % of the total bits is decreased by using the HoMG based DIC scheme in contrast to the DIC scheme in Fig. 4.28.

In general, the proposed HoMG based DIC scheme can achieve the most efficient network information coding in contrast to the original one. Performance comparison of these three schemes is shown in Fig. 4.29 by different color bars and curves. From the detail statistics in Table 4.4, approximately 10 and 64 % of the total bits are saved by using the DIC scheme and HoMG based DIC scheme against the original one, respectively. It can be concluded from the result that the HoMG based DIC scheme is the most efficient scheme for the broadcast CPC mode.

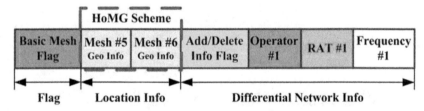

Fig. 4.26 Differential mesh frame format of HoMG based DIC scheme

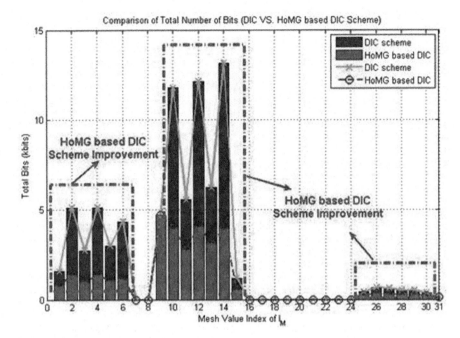

Fig. 4.27 Total bits comparison. (DIC scheme VS. HoMG based DIC scheme)

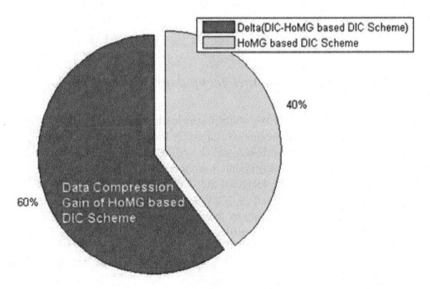

Fig. 4.28 Total bits proportion comparison. (DIC scheme VS. HoMG based DIC scheme)

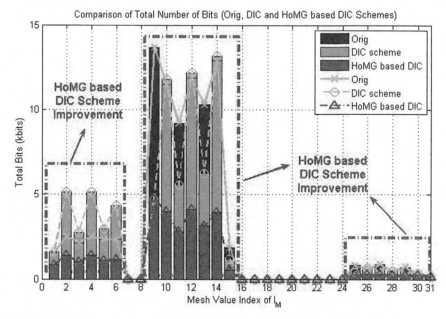

Fig. 4.29 Total bits comparison. (Original, DIC scheme and HoMG based DIC scheme)

Table 4.4 Total bits comparison (Orig, DIC, HoMG_DIC)

	Original	DIC	HoMG_DIC
Total bits (kbits)	88.62	79.82	31.67
Improvement	N/A	10%	64.26%

4.3.3 Cognitive Pilot Channel Technology Standard Evolution

As an efficient technology for the control information delivery in heterogeneous networks, CPC has proposed the novel solution for the separation of control signaling and data information transmission, which applies a large coverage network for frequently used information delivery and a small coverage network for the detailed information delivery in specific locations in [34, 36] in 2009.

Similar solutions are discussed in 3GPP standard from 2012 recently in [39], which has also paid much attention to the control signaling design by dual connectivity technology for data traffic offloading in heterogeneous network scenario for the small cell enhancement in LTE release 12. As proposed in [39], the dual connectivity technology utilizes the large coverage macro cell network for the connection and mobility control signaling and the small coverage femto cell network for data transmission. By using the separation of control signaling and data information delivery, the dual connectivity technology can enhance the load balance among macro and femto cells for both uplink and downlink, decrease the number of signaling overhead in frequently handovers and improve the capacity of heterogeneous networks.

Therefore, the key idea of CPC technology, which utilizes different networks for the control signaling and data information transmission by separation or decoupling of the control and data transmission in traditional networks, has been considered as a key solution for the enhancement of small cells in 3GPP [39]. Besides, CPC technology has also been accepted as a key technology by ETSI [41, 42] and ITU [43, 44].

4.4 Summary

This chapter focuses on the cognitive information obtained in CWN. In order to achieve the best overall sensing performance in coordinated spectrum sensing, a greedy algorithm and an iterative Hungarian algorithm are proposed to optimize the assignment of the SUs for multiple sensing channels. Simulation results show that the greed algorithm achieves satisfactory performance while the iterative Hungarian Algorithm based strategy is prior to that of the greedy based strategy. The geo-location database and multi-domain database are also introduced as efficient data storage schemes. Different from the geo-location database, the multi-domain database is proposed by adding new information domains. Besides, a novel scheme using three regions is proposed for spectrum sensing in the space-time domain. The closed-form bounds of three regions are obtained based on theoretical analysis, which can be used in the space-time spectrum sensing and access in cognitive wireless networks. Moreover, a HoMG based DIC scheme is proposed to achieve an efficient network information coding for the efficient cognitive information deliver using broadcast CPC. Performance analysis shows that approximately 10 % and 64 % of the total bits are saved by using the DIC scheme and HoMG based DIC scheme against the original one, respectively.

References

1. Haykin S, Thomson DJ, Reed JH (2008) Spectrum sensing for cognitive radio. Proceedings of the IEEE 97(5):849–877
2. Kolodzy P (2001) Next generation communications: kickoff meeting. Paper presented at the proceedings of DARPA, Oct 2001
3. Yucek T, Arslan H (2009) A survey of spectrum sensing algorithms for cognitive radio applications. IEEE Commun Surv Tutor 11(1):116–130
4. FCC Doc ET Docket: 03-289 (2003) Establishment of interference temperature metric to quantify and manage interference and to expand available unlicensed operation in certain fixed mobile and satellite frequency bands, FCC
5. Yucek T, Arslan H (2006) Spectrum characterization for opportunistic cognitive radio systems. Paper presented at the proceedings of military communication conference, Washington DC, 23–25 Oct 2006

6. Quan Z, Shellhammer SJ, Zhang W et al (2009) Spectrum sensing by cognitive radios at very low SNR. Paper presented at the proceedings of global communications conference, Beijing, Washington DC, 30 Nov 2009

7. Tang H (2005) Some physical layer issues of wide-band cognitive radio systems. Paper presented at the proceedings of international symposium on new frontiers in dynamic spectrum access networks, Baltimore, Washington DC, 8–11 Nov 2005

8. Weidling F, Datla D, Petty V et al (2005) A framework for RF spectrum measurements and analysis. Paper presented at the proceedings of international symposium on new frontiers in dynamic spectrum access networks, Baltimore, Washington DC, 8–11 Nov 2005

9. Lehtomaki L, Vartiainen J, Juntti M et al (2006) Spectrum sensing with forward methods. Paper presented at the Proceedings of MILCOM, Washington DC, 23–25 Oct 2006

10. Gardner UW (1991) Exploitation of spectral redundancy in cyclostationary signals. IEEE Sign Proces Mag 8(2):14–36

11. Muraoka K, Ariyoshi M, Fujii T (2008) A novel spectrum-sensing method based on maximum cyclic autocorrelation selection for cognitive radio system. Paper presented at the proceedings of 3rd symposium on new frontiers in dynamic spectrum access networks, Dublin, Washington DC, 14–17 Oct 2008

12. Du KL, Wai HM (2010) Affordable cyclo-stationarity-based spectrum sensing for cognitive radio with smart antennas. IEEE Trans Veh Technol 59(4):1877–1887

13. Proakis JG (2001) Digital communications, 4th edn. McGraw-Hill, New York

14. Tandra R, Sahai A (2005) Fundamental limits on detection in low SNR under noise uncertainty. Paper presented at the proceedings of international conference on wireless networks, communication and mobile computing, Maui, Washington DC, 13–16 June 2005

15. Tandra R, Sahai A (2008) SNR walls for signal detection. IEEE J Sel Top Sign Proces 2(1):4–17

16. Cabric D, Mishra S, Brodersen R (2004) Implementation issues in spectrum sensing for cognitive radios. Paper presented at the proceedings of asilomar conference on signals, system, computation, Washington DC, 7–10 Nov 2004

17. Cabric D, Tkachenko A, Brodersen R (2006) Spectrum sensing measurements of pilot, energy, and collaborative detection. Paper presented at the Proceedings of MILCOM, Washington DC, 23–25 Oct 2006

18. Hu W, Willkomm D, Abusubaih M et al (2007) Dynamic frequency hopping communities for efficient IEEE 802.22 operation. IEEE Commun Mag 45(5):80–87

19. Hillenbrand J, Weiss T, Jondral F (2005) Calculation of detection and false alarm probabilities in spectrum pooling systems. IEEE Commun Lett 9(4):349–351

20. Peh E, Liang YC (2007) Optimization of cooperative sensing for cognitive radio networks. Paper presented at the Proceedings of Wireless Communications and Networking Conference, Hong Kong, Washington DC, 11–15 March 2007

21. Liang YC, Zeng YH, Peh E et al (2008) Sensing-throughput tradeoff for cognitive radio networks. IEEE Trans Wireless Commun 7(4):1326–1337

22. Meng J, Yin WT, Li HS et al (2011) Collaborative spectrum sensing from sparse observation in cognitive radio networks. IEEE J Sel Areas Commun 29(2):327–337

23. Duan DL, Yang LQ, Principe JC (2010) Cooperative Diversity of Spectrum sensing for cognitive radio systems. IEEE Trans Sign Proces 58(6):3218–3227

24. Li HS, Zhu H (2010) Catch me if you can: an abnormality detection approach for collaborative spectrum sensing in cognitive radio networks. IEEE Trans Wireless Commun 9(11):3554–3565

25. Wang ZL, Feng ZY, Zhang P (2011) An iterative Hungarian algorithm based coordinated spectrum sensing strategy. IEEE Commun Lett 15(1):49–51

26. Lee CH, Wolf W (2007) Multiple access-inspired cooperative spectrum sensing for cognitive radio. Paper presented at the IEEE MILCOM, Orlando, FL, 29–31 Oct 2007

27. Kuhn HW (1955) The Hungarian Method for the Assignment and Transportation Problems. Naval Res Logist Quart 2(1–2):83–97

28. Bkassiny M, Jayaweera SK (2010) Optimal channel and power allocation for secondary users in cooperative cognitive radio networks. Paper presented at the special session on advanced radio access techniques for energy-efficient communications in 2nd international conference on mobile lightweight wireless systems, Barcelona, Spain, 10–12 May 2010

29. Federal Communications Commission (2008) In the matter of unlicensed operation in the tv broadcast bands: second report and order and memorandum opinion and order. Tech Rep

30. Kang KM, Park JC, Cho SI et al (2012) Deployment and coverage of cognitive radio networks in TV white space. IEEE Commun Lett 50(12):88–94

31. ECC report 159 (2011) Technical and operational requirements for the possible operation of cognitive radio systems in the 'White Spaces' of the frequency band 470–790 MHz, report out for public consultation

32. Wang ZL, Feng ZY, Zhang D et al (2011) optimized strategies for coordinated spectrum sensing in cognitive radio networks. Paper presented at the CROWNCOM, Osaka, 1–3 June 2011

33. IST-2003-507995 E2R (End-to-End Reconfigurability) Project, http://e2r.motlabs.com

34. Zhang QX, Feng ZY (2009) A novel mesh division scheme using cognitive pilot channel in cognitive radio environment. Paper presented at the IEEE 70th VTC, Anchorage, 20–23 Sept 2009

35. Zhang QX, Feng ZY, Zhang GY (2010) A novel homogeneous mesh grouping scheme for broadcast cognitive pilot channel in cognitive wireless networks. Paper presented at the IEEE International Conference on Commun, Cape Town, 23–27 May 2010

36. Zhang QX, Feng ZY, Zhang GY (2012) Efficient mesh division and differential information coding schemes in broadcast cognitive pilot channel. Wireless Pers Commun 63(2):363–392

37. Wei ZQ, Feng ZY (2011) A geographically homogeneous mesh grouping scheme for broadcast cognitive pilot channel in heterogeneous wireless networks. Paper presented at the IEEE GLOBECOM Workshops, Houston, 5–9 Dec 2011

38. Feng ZY, Wei ZQ, Zhang QX et al (2012) Fractal theory based dynamic mesh grouping scheme for efficient cognitive pilot channel design. Chin Sci Bull 57(28–29):3684–3690

39. 3GPP TR 36.932 V12.1.0 (2013) Scenarios and requirements for small cell enhancements for E-UTRA and E-UTRAN (Release 12)

40. Perez-Romero J, Sallent O, Agusti R et al (2007) A novel on-demand cognitive pilot channel enabling dynamic spectrum allocation. Paper presented at the 2nd IEEE International Symposium on New Frontiers in Dynamic Spectrum Access Networks, Dyspan, 17–20 April 2007

41. ETSI TR 102 683 (2009) Reconfigurable radio systems (RRS)—cognitive pilot channel (CPC)

42. ETSI TR 102 682 (2009) Reconfigurable radio systems (RRS)—functional architecture for management and control of reconfigurable radio systems

43. Report ITU-R M.2225 (2011) Introduction to cognitive radio systems in the land mobile service

44. Report ITU-R M.2242 (2011) Cognitive radio systems specific for international mobile telecommunications systems

Chapter 5
Intelligent Resource Management

The major challenges, faced by modern wireless communications, originate from the fluctuating feature of the available spectrum, as well as the diverse QoS requirements of various applications. Intelligent resource management can address these challenges to realize a new network paradigm. The traditional radio resource management (RRM) is applied to provide qualified service for wireless terminals to fully improve the utilization of wireless resources by dynamically allocating and adjusting the available network resources, especially in the situation where network traffic is unevenly distributed and the channel characteristics fluctuate due to channel fading and random interference. The RRM includes admission control, channel allocation, switching, load balancing, packet scheduling, power control and so on. However, future challenges are the coexistence of various heterogeneous wireless networks, which will lead to the traditional RRM inferior. So, Dynamic Spectrum Management (DSM), Joint Radio Resource Management (JRRM) and Dynamic Spectrum Access (DSA) are proposed.

5.1 Dynamic Spectrum Management

The mechanisms and the structure of DSM are shown in Fig. 5.1. The network should learn and know the environment in the first place. And then based on the learned information, the network handles the calculation of the dynamic spectrum management to achieve the best spectrum allocation strategy. DSM applies the basic four steps of cognitive theory. They are observation, cognition, decision making and implementation process. Thus DSM can optimize the use of spectrum resources for isomorphic and heterogeneous networks, single-operator and multi-operator.

Different from the traditional static spectrum allocation, DSM effectively and dynamically allocates spectrum for the RATs by utilizing the load differences in time and space dimensions of different RATs, which means spectrum no longer belongs to one RAT permanently. Thus DSM improves the utilization of spectrum and system capacity.

© The Author(s) 2015   85
Z. Feng et al., *Cognitive Wireless Networks,* SpringerBriefs in Electrical
and Computer Engineering, DOI 10.1007/978-3-319-15768-9_5

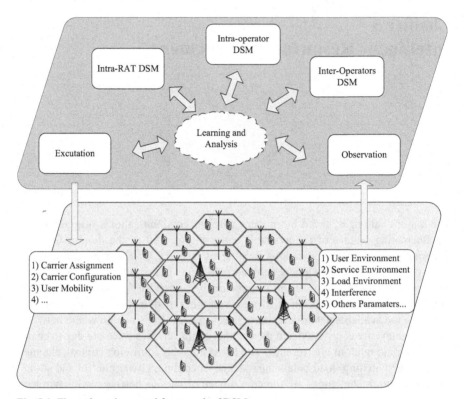

Fig. 5.1 The active scheme and framework of DSM

Dynamic Spectrum Management can be divided as follows.

- Dynamic Spectrum Sharing (DSS): DSS allocates the spectrum to multiple networks at the same time to achieve spectrum sharing in data packet level, which can also be used to avoid conflicts.
- Dynamic Spectrum Allocation (DSA): Compared with DSS, DSA allocates the spectrum depending on users' requirements, which belongs to the long-term spectrum allocation.

Therefore, this book mainly concerns the long-term dynamic allocation of the authorized band among cellular networks. This section focuses on the dynamic spectrum allocation technology in heterogeneous networks.

5.1.1 System Model

Many researches [1–4] have proposed simple mechanism and process of DSM. As shown in Fig. 5.2, the process of DSM is made up of 4 basic steps.

Step 1, the estimation of traffic volume;

Step 2, the prediction of spectrum requirements;

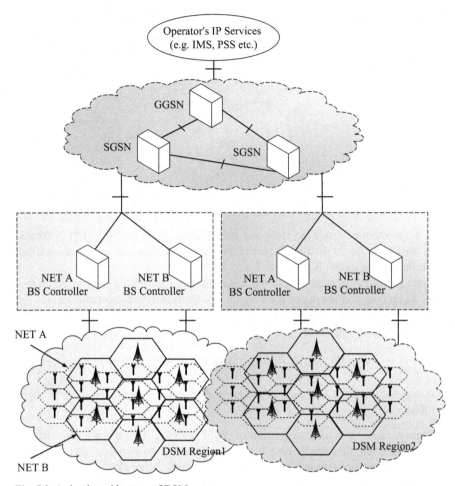

Fig. 5.2 A simple architecture of DSM

Step 3, run DSM algorithms;

Step 4, the configuration and use of spectrum.

As DSM can accustom to the differences of traffic requirements, network capacity and spectrum resources in heterogeneous networks, it not only reduces the spectrum holes, but also greatly improves both the utilization of spectrum resource and the capacity of heterogeneous systems at the same time.

5.1.2   Game Theory in DSM

In DSM, the spectrum trading is modeled as a dynamic game with the complete information. To solve this problem, the famous R-S (Rubinstein–Stahl) [5] game theory of microeconomics is utilized. Assuming that renter spectrum broker (SB)

Fig. 5.3 The DSM game
model

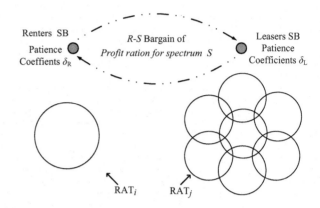

rents a spectrum of S (Hz) to lease and gain a profit of P, as the spectrum comes
from leaser, the leaser should get part of the profit P. Using R-S theory model, the
spectrum trading model is shown in Fig. 5.3.

Taking profit P as a cake, both the renters and the leasers bargain based on the
cake. Assuming X is the set of the bargain results, results are shown in Eq. (5.1),
where x_R and x_L refer to the ratio of profit that renters and leasers get correspondingly.

$$X = \{(x_R, x_L) \in R^2 : x_R \geq 0, x_L \geq 0, and, x_R + x_L = 1\} \qquad (5.1)$$

In R-S game theory, the process of the bargaining can be concluded in three steps
as follows.

1. Renters come up with a suggestion of profit ratio (x_R^1, x_L^1). If the leasers accept
   the suggestion, then the bargain is finished; If not, go to step 2.
2. Leasers come up with another suggestion of profit ratio (x_R^2, x_L^2). If the renters
   accept the suggestion, then the bargain is finished; If not, go to the next step.
3. Step 1 and step 2 will take actions in turns until renters and leasers reach a
   consensus.

R-S theory model assumes that both the renters and the leasers have limited patience and the patience coefficients of the renters and the leasers are defined as
δ_R and δ_L correspondingly. It is obvious that more patience one side has, the more
profit ratio it will get at last.

Therefore, the patience coefficients depend on the spectrum value to two sides of
the game. Thus, in order to acquire the results of the game, the spectrum economic
value of different game sides needs to be estimated and their patience coefficients
need to be calculated thereafter.

The Estimation of Spectrum Economic Value
Assuming that the total spectrum of an RAT is B and the bandwidth is S. Then the
estimation of the spectrum value is denoted in Eq. (5.2).

$$V^S = \frac{L\alpha}{B}S \tag{5.2}$$

The Patience Coefficients of Renters
As to spectrum renters, the spectrum value reflects the extent of their need for spectrum. The patience of renters is low because they need spectrum, resulting in that the renters' patience coefficients is a decreasing function of the economic value of spectrum. In R-S game model, patience coefficients have a value from 0 to 1[5]. For renters, the patience of the renters should satisfy Eq. (5.3).

$$\frac{d\,\delta_R(V_{Rent}^S)}{dV_{Rent}^S} < 0,\ \delta_R(0) = 1,\ \delta_R(\infty) = 0 \tag{5.3}$$

V_{Rent}^S is the economic value estimation of S (Hz) spectrum.

Theoretically, any function that satisfies the above conditions can be taken as the patience factor function, by using Eq. (5.4) to acquire the patience factor [6].

$$\delta_R(x) = 1 - \frac{e^{\lambda x} - e^{-\lambda x}}{e^{\lambda x} + e^{-\lambda x}} \tag{5.4}$$

x is the economic value of spectrum for renters and λ is the lease factor.

The Patience Coefficients of Leasers
As to leasers, the higher the spectrum value is, the more profits they can acquire, which means that their patience coefficients are higher. Thus, a patience coefficient is an increasing function of the economic value of spectrum for leasers. In R-S model, the patience coefficients of the leasers should satisfy Eq. (5.5).

$$\frac{d\,\delta_L(V_{Lease}^S)}{dV_{Lease}^S} > 0,\ \delta_L(0) = 1,\ \delta_L(\infty) = 0 \tag{5.5}$$

V_{Lease}^S is the spectrum economic value for leasers. In addition that if the spectrum covers several RATs, V_{Lease}^S is the total spectrum value of all the RATs as different RATs will have different economic value.

Then patience coefficients of leasers can be expressed in Eq. (5.6).

$$\delta_L(x) = \frac{e^{\mu x} - e^{-\mu x}}{e^{\mu x} + e^{-\mu x}} \tag{5.6}$$

x is the spectrum economic value of leasers and μ is the lease factor.

In spectrum trading, both the spectrum renters and leasers have a game based on the profit, which can achieve a Nash Equilibrium [5]. When Nash Equilibrium is achieved, the game results are denoted in Eq. (5.7).

$$(x_R^*, x_L^*) = \left(\frac{1-\delta_L}{1-\delta_R\delta_L},\ \frac{\delta_R(1-\delta_L)}{1-\delta_R\delta_L} \right) \tag{5.7}$$

In addition, as the game is based on complete information, so in order to minimize signaling and network processing cost, in practical DSM model the game process will directly calculate the Nash Equilibrium results but not bargaining via SB messages of practical DSM model. At the end of each DSM cycle, the profit is allocated and the renters will return the spectrum rented.

5.1.3   Performance Analysis and Evaluation

The simulation scenario contains 4 heterogeneous networks, in which both GSM and UMTS have 14 RATs and DVB-T has 1 RAT. The spectrum bandwidth of GSM, UMTS and DVB-T is 7, 15 and 24 (MHz) correspondingly. In addition, the area of each A_{pq} is set by 2500 m^2. The user density δ_{pq} of A_{pq} is generated randomly and the variance σ_{pq} is randomly selected in $\{4,6\}$ as denoted in [7].

The path loss model is applied as denoted in Eq. (5.8) by 3GPP 25.942 in [8].

$$Los(dB) = 15.3 + 36.7 \lg(d(m)) \tag{5.8}$$

For the traffic prediction, the technique that combines historical data prediction and linear regression prediction based on two samples is adopted in [9]. According to the measurements, the traffic volume of GSM and UMTS is high in day time and low at night. However, for DVB-T, the results are on the contrary. It means that GSM/UMTS and DVB-T have a good complementary for each other and the differences of time dimension can fully be utilized to improve the spectrum utilization and network profit. Figures 5.4, 5.5, and 5.6 show the volume distribution of the heterogeneous networks in one day. Figures 5.7, 5.8, 5.9, and 5.10 show the network profit based on. Figures 5.11, 5.12, and 5.13 show the distribution curve of interference probability.

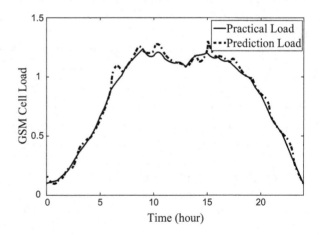

Fig. 5.4 GSM volume prediction

Fig. 5.5 UMTS volume prediction

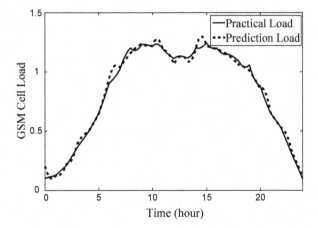

Fig. 5.6 DVB volume prediction

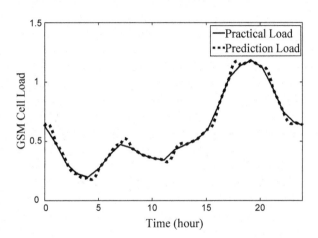

Fig. 5.7 GSM1 network profit

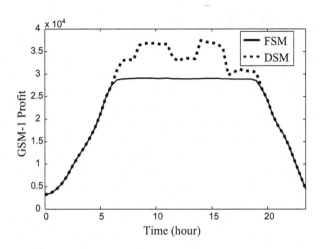

Fig. 5.8 GSM2 network
profit

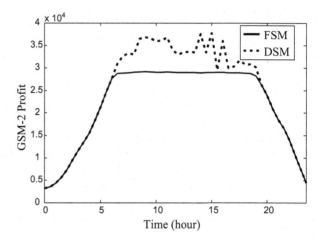

Fig. 5.9 UMTS network
profit

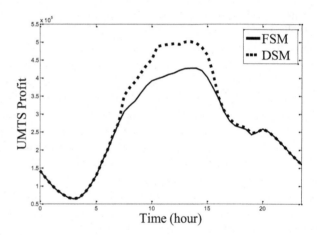

Fig. 5.10 DVB network
profit

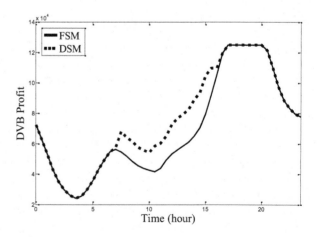

Fig. 5.11 GSM1 interference
distribution

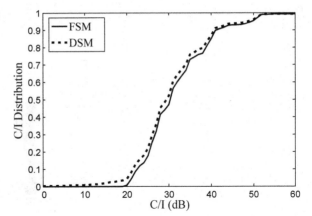

Fig. 5.12 GSM2 interference
distribution

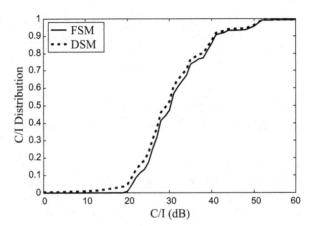

Fig. 5.13 UMTS interference
distribution

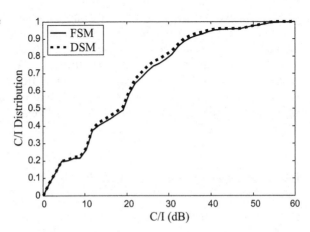

Fig. 5.14 Spectrum utilization

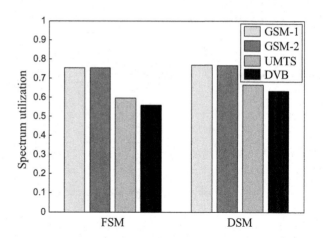

In Fig. 5.14, a high traffic volume for GSM_1 and GSM_2 is set but a low traffic volume for DVB is set in order to make simulation closer to the actual situation. Thus the spectrum utilization improvement of GSM is low but DVB gets a huge improvement in spectrum utilization.

5.2 Joint Radio Resource Management

Joint Radio Resource Management (JRRM) is an evolution of the existing RRM. It is also a set of network control mechanisms, which considers multiple wireless networks comprehensively based on all RRM functions. Moreover, JRRM is able to support intelligent call and session admission control as well as distributed processing of services and power. Thus, optimization of the radio resources and maximization of the system capacity can be achieved.

As the network control mechanism aiming at optimizing the utilization of the radio resources and maximizing the system capacity, JRRM is capable of supporting intelligent voice service request, session admission control and distribution of service flow as well as power between different RATs. Joint Session Admission Control (JOSAC) [10] is an important method of JRRM. In this section, JOSAC will be elaborated and multi-operator scenario will be considered. The other two single-operator scenarios, which depict the distributed autonomic JOSAC and centralized autonomic JOSAC, can be referred in [11, 12].

5.2.1 Reinforcement Learning in JRRM

A standard RL problem can be modeled as Markov decision process (MDP) $< S, A, R, T$, where $S = \{s_1, s_2, ..., s_n\}$ is the possible state space of the environment, $A = \{a_1, a_2, ..., a_m\}$ is the possible action space of the agent, $R : S \times A \rightarrow PD(S)$ is the

Fig. 5.11 GSM1 interference
distribution

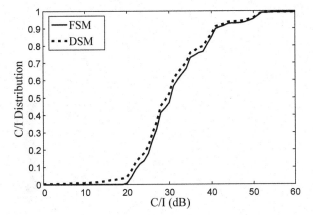

Fig. 5.12 GSM2 interference
distribution

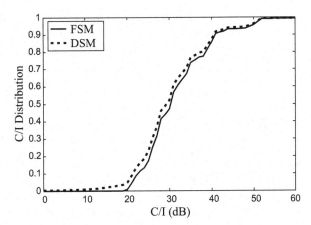

Fig. 5.13 UMTS interference
distribution

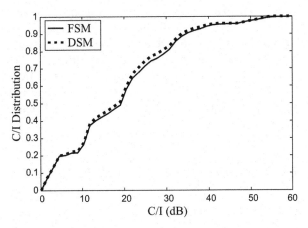

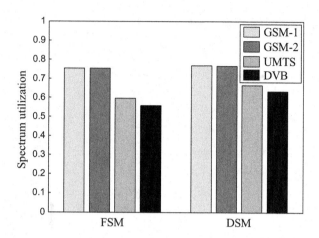

Fig. 5.14 Spectrum utilization

In Fig. 5.14, a high traffic volume for GSM_1 and GSM_2 is set but a low traffic volume for DVB is set in order to make simulation closer to the actual situation. Thus the spectrum utilization improvement of GSM is low but DVB gets a huge improvement in spectrum utilization.

5.2   Joint Radio Resource Management

Joint Radio Resource Management (JRRM) is an evolution of the existing RRM. It is also a set of network control mechanisms, which considers multiple wireless networks comprehensively based on all RRM functions. Moreover, JRRM is able to support intelligent call and session admission control as well as distributed processing of services and power. Thus, optimization of the radio resources and maximization of the system capacity can be achieved.

As the network control mechanism aiming at optimizing the utilization of the radio resources and maximizing the system capacity, JRRM is capable of supporting intelligent voice service request, session admission control and distribution of service flow as well as power between different RATs. Joint Session Admission Control (JOSAC) [10] is an important method of JRRM. In this section, JOSAC will be elaborated and multi-operator scenario will be considered. The other two single-operator scenarios, which depict the distributed autonomic JOSAC and centralized autonomic JOSAC, can be referred in [11, 12].

5.2.1   Reinforcement Learning in JRRM

A standard RL problem can be modeled as Markov decision process (MDP) $< S, A, R, T$, where $S = \{s_1, s_2, ..., s_n\}$ is the possible state space of the environment, $A = \{a_1, a_2, ..., a_m\}$ is the possible action space of the agent, $R : S \times A \to PD(S)$ is the

Fig. 5.15 The standard RL
model

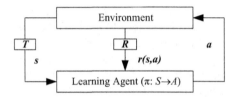

reward function of the agent, and $T : S \times A \rightarrow PD(S)$ is the state transition function,
where $PD(S)$ denotes the set of probability distributions over S.

The basic RL model is illustrated in Fig. 5.15. The agent perceives the state $s \in S$
of the environment and decides the action $a \in A$ following its current policy $\pi S \rightarrow A$.
Consequently, the environment may change to a state $s' \in S$ according to T and the
agent receives a scalar reinforcement signal $r(s,a)$, called the immediate reward,
according to R. Then the agent updates its policy using s' and $r(s,a)$. Such a process continues in an iterative way and the final goal is to find the optimal policy
$\pi^*(s) \in A$ that maximizes the agent's expected long term reward (or the value) in
each state as denoted in Eq. (5.9).

$$V^\pi(s) = E\left(\sum_{t=0}^{\infty} \gamma^t r(s_t, \pi(s_t)) \mid s_0 = s \right),   \tag{5.9}$$

$\gamma \in (0,1)$ is the discount factor reflecting the significance of the future reward rela-
tive to the current one.

Single-agent RL has been well studied during the past decades. One of the most
popular algorithms is Q-learning [13], It associates a Q-value with each pair of
state-action (s, a) and learns the optimal policy through the simple *value-iteration*
rule in Eqs. (5.10) and (5.11).

$$Q_{t+1}(s,a) = (1-\alpha)Q_t(s,a) + \alpha(r_t + \gamma V(s'))   \tag{5.10}$$

$$V(s) = \max_{a' \in A} Q_t(s,a')   \tag{5.11}$$

$\alpha \in [0,1)$ is the learning rate. As $t \rightarrow \infty$, if the learning rate is decreased suitably to 0
and the Q-value of each (s,a) pair is visited infinitely frequently, $Q_t(s,a)$ converges
to the optimal value $Q^*(s,a)$ with probability 1 in [13]. Then, the optimal policy is
obtained as denoted in Eq. (5.12).

$$\pi^*(s) = \arg\max_a Q^*(s,a)   \tag{5.12}$$

In a multi-agent context, the state transition T is no longer guaranteed to be *Mar-
kovian* from a single agent's point of view, since its environment is affected by
other agents' actions. Thus the direct application of traditional RL algorithms (e.g.
Q-learning) in a multi-agent environment could be problematic. By extending each

state of a MDP as a *matrix game* between agents, a *Markov game* framework has been developed for MARL and a handful of algorithms [14–17] are now available based on the Q-learning principle. The common idea includes the extension of individual action space to the joint action space $A = A_1 \times A_2 \times ... \times A_k$ where A_i is the ith agent's action space, and the use of Q-values as the agents' payoff matrices to decide the joint action (or equilibrium) in each state. The differences lie in the target problem types and the methods to find the solution.

The algorithm can be implemented by replacing Eq. (5.11) in Q-learning with following linear programming problem as denoted in Eq. (5.13).

$$V_i(s) = \max_{\pi_i \in PD(A_i)} \min_{a_{-i} \in A_{-i}} \sum_{a_i \in A_i} Q_i(s, a_i, a_{-i}) \pi_i(s, a_i)$$

$$s.t. \quad \sum_{a_i \in A_i} \pi_i(s, a_i) = 1 \quad \forall i = 1, 2 \tag{5.13}$$

$$0 \le \pi_i(s, a_i) \le 1 \quad \forall i = 1, 2$$

a_{-i} and A_{-i} denote respectively the action and action space of the ith agent's opponent. The solution to Eq. (5.13) is a probabilistic policy (i.e. a mixed strategy in game theory) instead of the deterministic one in the traditional Q-learning.

With the joint action extension $a \in A$, correlation Q-learning (CE-Q) can be implemented based on Q-learning by replacing Eq. (5.11) with the following value function in Eq. (5.14).

$$V_i(s) = \sum_{a \in A} Q_i(s, \mathbf{a}) \pi^*(s, \mathbf{a}) \tag{5.14}$$

$\pi^*(s, a)$ is the mixed strategy at CE and can be achieved as the solution to the linear programming problem in Eq. (5.15).

$$\max_{\pi \in PD(A)} \sum_i \sum_{a \in A} Q_i(s, \mathbf{a}) \pi(s, \mathbf{a})$$

$$s.t. \quad \sum_{a_{-i} \in A_{-i}} (Q_i(s, a_i, \mathbf{a}_{-i}) - Q_i(s, a_i', \mathbf{a}_{-i})) \pi(a_i, \mathbf{a}_{-i}) \ge 0$$

$$\forall a_i, a_i' \in A_i \text{ for all } i,$$

$$\sum_{a \in A} \pi(s, \mathbf{a}) = 1$$

$$0 \le \pi(s, \mathbf{a}) \le 1 \qquad \forall \mathbf{a} \in A, \tag{5.15}$$

a_i and a_j' denote two different action within the ith agent's action space and a_{-i} denotes other agents' joint action.

Considering the multi-agent scenario, the CE-Q algorithm with the Markov game formulation is proposed for JRRM. The current implementation is for a 2-agent case, but can also be extended to the n-agent case using multi-dimensional Q-value matrices as in n-player general-sum games. Before introducing the working flow of the proposed algorithm, the system architecture is proposed first with the consideration of the input generalization.

5.2.2   *Performance Analysis and Evaluation*

In all algorithms, the three layers MFNNs are adopted with hyperbolic tangent and linear transfer functions for the hidden layer and output layer respectively. The number of the hidden-layer nodes is chosen empirically according to [18], in order to satisfy the approximation requirement. The use of MFNN saves the memory space to a great extent.

Fig. 5.16 Blocking probability under different traffic conditions

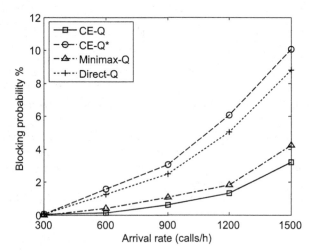

Fig. 5.17 Total network profits under different traffic conditions

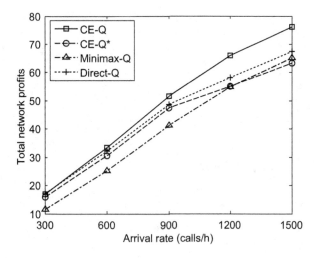

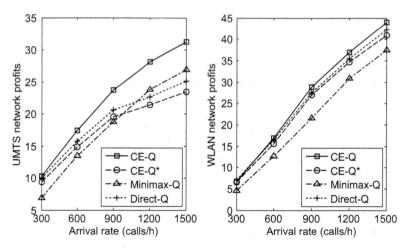

Fig. 5.18 Profits in each network under different traffic conditions

First, the overall blocking probability and total network profits under different traffic conditions are shown in Figs. 5.16 and 5.17. The network profits are calculated as the sum of the reward of serving each session in the network. All results are averaged over 50 runs.

It's also shown that the information sharing could be reciprocal and self-motivated as manifested by the results in Fig. 5.18, where the respective profits of each network can be improved simultaneously if they both choose CE-Q.

Meanwhile, different types of services are prioritized in each network as shown in Fig. 5.19. RT and NRT services account for the major traffic portion in UMTS

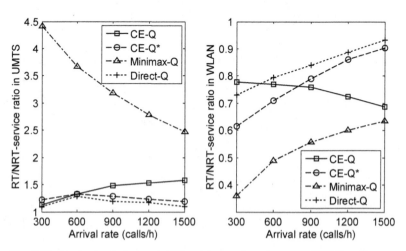

Fig. 5.19 The load ratio between different types of services (RT/NRT) in each network under different traffic conditions

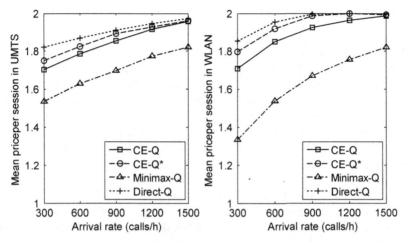

Fig. 5.20 Mean price charged per session in each network under different traffic conditions

and WLAN respectively, thus the limited resources are utilized efficiently to produce higher profits. However, the ratio varies differently under each algorithm and only CE-Q shows the desirable trend, i.e. services are more differentiated when systems are highly loaded. In fact, as shown in Fig. 5.20, the pricing policy of Minimax-Q is quite conservative compared to CE-Q because it always wants to assure the worst-case profits by attracting the users with lower price. Due to the correlation at the equilibrium, CE-Q achieves the highest profits by more aggressive pricing policies of the agents.

5.3 Dynamic Spectrum Access

DSA has become a promising technique to fully utilize the scarce spectrum resources. However, spectrum allocation schemes with high efficiency and quality of service (QoS) guaranteed for primary users have drawn much attention recently. Therefore, two novel DSA schemes based on Markov chains are proposed to efficiently share the licensed spectrum among secondary users. First, two DSA schemes in the absence and presence of buffer are proposed for secondary users to make use of the licensed spectrum. Second, the system's evolutionary behavior, including the primary user's activities, is thoroughly captured through continuous-time Markov chain (CTMC) modeling. Third, the proposed schemes take the effects of spectrum sensing errors into consideration for the optimal performance. By deriving the optimal access probabilities for the secondary users, the total throughput of secondary system is maximized with QoS guaranteed for primary users.

5.3.1   System Model

A dynamic spectrum access network is considered that multiple secondary users are allowed to access the temporarily unused licensed spectrum bands in an
opportunistic way. In particular, the scenario is considered where M secondary users
coexist with one primary user, and secondary users contend to access to the vacant
spectrum. The primary user is denoted as P, which owns the license of the spectrum
band. For the primary user, the arrival traffic is modeled as a Poisson process with
rate λ_p. The radio system access duration is negative-exponentially distributed with
a mean time $1/\lambda\mu$, hence, the departure of user P's traffic is another Poisson process
with rate $\lambda\mu$. The secondary user's traffic is also modeled as two Poisson random
processes, with arrival rate λ_i and departure rate μ_i, respectively ($i = 1, 2, ..., M$).

To achieve the maximum utilization of the spectrum without causing extra interference to the primary user, it is assumed that the secondary users are capable of
detecting the primary user's activities with spectrum sensors, i.e., the arrival and
departure of primary user in the spectrum band. It is also assumed that the secondary users are not permitted to share the spectrum band simultaneously because the
interference would result in throughput degradation. The secondary users' access is
controlled by a secondary central node C. All secondary users should report their
access activities to node C, i.e., if secondary user i has successfully accessed to
the spectrum band, the central node C should be informed to mark the band as allocated. Furthermore, imperfect spectrum sensing which consists of false alarm and
miss-detection is considered. The former occurs when an idle channel is detected as
busy, and the latter occurs when a busy channel is detected as idle. The probabilities
of false alarm and miss-detection are denoted by P_f and P_d respectively.

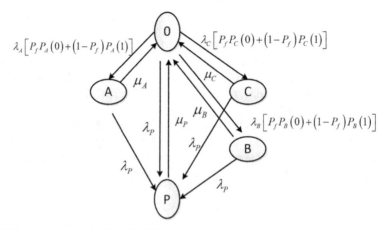

Fig. 5.21 System state transition for DSA

5.3.2 DSA Schemes Based on Markov Theory

In this scheme, the collisions among secondary users are ignored because this type of collisions only occur when their service requests arrive exactly at the same time. This can be neglected for services following the dependent Poisson process. Since there is no buffering mechanism for the secondary users, a simple Markov chain is developed. Figure 5.21 illustrates an example of this scheme, where there are three secondary users (A, B, C) and one primary user P. State 0 means the spectrum band is idle and state S means user S operates in the band, with $S \in \{A, B, C, P\}$.

The collision between the primary and secondary users may happen due to spectrum sensing errors. As the probability of miss-detection increases, the collision probability also increases. In order to guarantee a good protection for the primary user, secondary users' access should be strictly controlled. In the proposed scheme, the primary network collision constraints are guaranteed as well as refraining secondary network from overlooking too many opportunities by introducing the spectrum access probabilities. In the scheme, secondary user i is associated with two probabilities $P_i(0)$ and $P_i(1)$, which are obtained by the central node C previously described.

When user i's traffic arrives, node i will access the spectrum band randomly with probability $P_i(1)$ if it detects the band as idle. On the other hand, if i detects the band as busy, it will access the spectrum band randomly with probability $P_i(0)$, therefore, unused spectrum opportunities resulted from false alarm can be utilized by secondary users. Since secondary user's access, e.g., A, is controlled by the two probabilities $P_i(0)$ and $P_A(1)$, the rate that the chain transits from state 0 to state A can be obtained as $\lambda_A \left[P_f P_A(0) + (1 - P_f) P_A(1) \right]$. Once user A's service is completed before user P requests spectrum access, the chain transits from state A to state 0 with rate μ_A. To guarantee primary user P's priority, a secondary user's traffic should be promptly dropped or switched to anther band when user P appears during the service duration of the secondary user. It is also assumed that when primary system is about to reclaim the band, it would send a strong alert signal in advance, so that the secondary user is able to detect this alert signal correctly and releases the channel promptly.

The "flow-balance" (the rate transitions out of state S equals to the rate transitions into state S) and the normalization equations governing the above system are denoted in Eq. (5.16).

$$\mathbf{H} \, \pi^{\mathrm{T}} = 0 \qquad\qquad (5.16)$$

$\mathbf{H}$ is the matrix that characterizes the transition states of the Markov chain, and $\pi = \left[\pi_0, \pi_1, \ \dots, \ \pi_M, \pi_P \right]$ is the steady state probability vector. By solving Eqs. (5.1) and (5.2), the steady-state probabilities are obtained as shown in Eq. (5.17).

$$\pi_0 = \mu_P C^{-1} \prod_{i=1}^{M} (\mu_i + \lambda_P) \tag{5.17}$$

It is assumed that the transmission rate of secondary user i is r kbps. Therefore, the total average throughput for user i acquired on the licensed band is depicted in Eq. (5.18).

$$R_i = \lim_{T \to \infty} \frac{1}{T} \int_0^T r \Pr\{S(t) = 1\} dt$$
$$= r \lim_{T \to \infty} \frac{1}{T} \int_0^T \Pr\{S(t) = 1\} dt \tag{5.18}$$
$$= r \pi_i$$

The interference to primary user on MAC layer should be limited under a certain target value. As mentioned above, the random access policy for secondary users is applied, through which the QoS of primary user in terms of collision probability constraint is guaranteed. In the duration of primary user P's service, the average number of traffic requests for user i can be given as $n_i = \lambda_i / \mu_P$. The collision probability between primary and secondary users can be expressed in Eq. (5.19).

$$P_c = 1 - \prod_{i=1}^{M} (1 - [(1 - P_d)P_i(0) + P_d P_i(1)])^{\frac{\lambda_i}{\mu_P}} \tag{5.19}$$

The goal is to determine $P_i(0)$ and $P_A(1)$, ($i = 1, 2, \ldots, M$), such that the system performance can be maximized. The optimal probabilities can be obtained by solving the following problem in Eq. (5.20).

$$\{\ldots P_i^{opt}(0), P_i^{opt}(1), \ldots\}$$
$$= \arg \max U(\ldots P_i(0), P_i(1), \ldots) \tag{5.20}$$

subject to

$$P_c \le P_{th} \tag{5.21}$$

The problem above can be transformed into a convex optimization problem by a transformation of the objective and constraint functions. The objective and constraint functions can be transformed into the following expressions in Eqs. (5.22) and (5.23).

$$U(P(0), P(1)) = \mu_P - \frac{\mu_P R}{R + g(P(0), P(1))} \tag{5.22}$$

$$\sum_{i=1}^{M} \frac{\lambda_i}{\mu_P} \ln\left(1-[(1-P_d)P_i(0)+P_d P_i(1)\right) - \ln(1-P_c^{th}) > 0 \qquad (5.23)$$

5.3.3   *Performance Analysis and Evaluation*

Figure 5.22 shows the optimal access probabilities with different probabilities of false alarm P_f and miss-detection P_d. In the left subplot, as P_d increases, the protection for primary user becomes more rigid, hence, $P_i(0)$ and $P_i(1)$ for each secondary user decrease. It is demonstrated through the simulation that the collision probability P_c keeps a constant value of 0.09 as P_d and P_f change. Therefore, the proposed scheme gives a good protection for the primary user. As can be seen from the right subplot, $P_i(0)$ is larger than $P_i(1)$ as P_f gets close to 1. This attributes to the fact that to access the band when detected busy is more beneficial than idle in terms of maximizing secondary system's throughput.

Figure 5.23 shows the total throughput of secondary users as P_d and P_f change. Moreover, the total throughput decreases as P_d and P_f increase. This observation reflects the essence of the randomized access policy to protect the QoS of primary users at the expense of sacrificing the throughput of the secondary system. It is also observed that as P_f gets close to 1, the total throughput increases slightly as P_f increases. The reason behind this observation is that the probability $P_i(0)$ increases

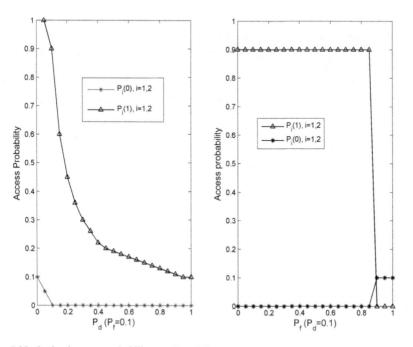

Fig. 5.22 Optimal access probability vs. P_d and P_f

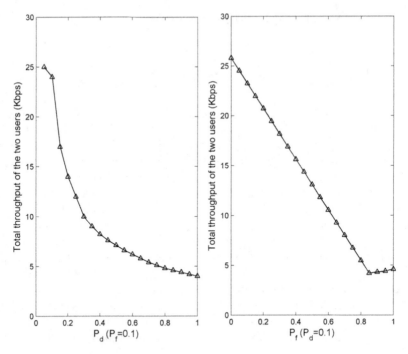

Fig. 5.23 Total throughput vs. P_d and P_f

according to the proposed maximal throughput criterion, and some unused spectrum opportunities resulted from false alarm are utilized by secondary users.

Fig. 5.24 shows the variation of the total throughput as the number of secondary user changes. The total throughput increases as the number of secondary user increases. However, the throughput remains constant when the number is larger than

Fig. 5.24 Total throughput
vs. number of secondary
users

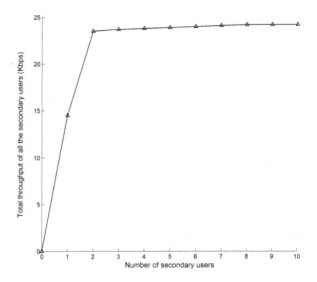

two. This is because the "spectrum hole" is filled up when the number grows large. The figure shows that the proposed scheme can provide about 23 kbps throughput for the secondary system on the band studied, which will be totally wasted without DSA scheme.

5.4 Concluding Remarks

In this chapter, challenges and solutions are described in detail on the intelligent radio resource management in CWNs. In terms of the dynamic changing features of users' services and unevenly distributed traffic in heterogeneous networks, dynamic spectrum management (DSM) scheme is proposed to improve the efficiency of spectrum management. Both the system model and game theory for DSM are proposed and verified by numerical results. Furthermore, to improve the efficiency of radio resource management, the joint radio resource management (JRRM) scheme is proposed, which is able to support voice service request, session admission control as well as distributed processing of services. To improve the spectrum efficiency and utilize vacant spectrum holes, dynamic spectrum access (DSM) technology is described with system model and Markov chains, which can efficiently share the licensed spectrum among secondary users. And the optimal access probabilities for secondary users are derived, which can maximize the total throughput of secondary system with QoS guaranteed for primary users.

References

1. IST-2005-027714 Project E2R II. End-to-End Reconfigurability phase 2. http://e2r2.motlabs. com/.
2. FP7-ICT-2007-216248 Project (2009) End-to-End Efficiency. Eurescom. http://www.ict-e3.eu. Accessed 7 Dec 2009
3. Leaves P, Moessner K, Tafazolli R et al (2004) Dynamic spectrum allocation in composite reconfigurable wireless networks. IEEE Commun Mag 42(5):72–81
4. Grandblaise D, Moessner K, Leaves P et al (2003) Reconfigurability support for dynamic spectrum allocation: from the DSA concept to implementation. Paper presented at the Mobile Future and Symposium on Trends in Communications, 26–28 Oct 2003
5. Rubinstein A, Osborne M (1990) Bargaining and markets. Academic Press, California, pp 29–49
6. Jeffrey A (2005) Mathematics for engineers and scientists, 6th edn. CRC Press, London, pp 321–323
7. Obeldobel J, Ruse H, Graf F (1996) Statistical modeling of time variant processes for interference analysis in mobile radio network planning. Paper presented at the International Symposium on Personal, Indoor and Mobile Radio Communications, Taiwan, Oct 1996
8. 3GPP TR 25.942 V7.0.0 (2007) Radio frequency (RF) system scenarios (Release 7)
9. Leaves P, Ghaheri-Niri S, Tafazolli R et al (2002) Dynamic spectrum allocation in hybrid networks with imperfect load prediction. Paper presented at the 3rd International Conference on 3G Mobile Communication Technologies, 8–10 May 2002
10. E2R Deliverable D5.3. Algorithms and Performance, including FSM & RRM/Network Planning (2005) http://e2r.motlabs.com

11. Zhang Y, Tang T, Chen J et al (2007) Autonomic joint session admission control using rein-forcement learning. J Beijing Univ Posts Telecommun 30(4):5–9
12. Zhang Y, Feng ZY, Zhang P et al (2008) A Q-learning based autonomic joint radio resource management algorithm. J Electr Info Technol 30(3):676–680
13. Watkins CJCH, Dayan P (1992) Q-learning. Mach Learn 8(3):279–292
14. Claus C, Boutilier C (1998) The dynamics of reinforcement learning in cooperative multia-gent systems. Paper presented at the Proceedings of the 15th National Conference on Artifi-cial Intelligence, Wisconsin
15. Littman ML (1994) Markov games as a framework for multi-agent reinforcement learning. Paper presented at the Proceedings of the 11th International Conference on Machine Learning
16. Hu J, Wellman MP (1998) Multiagent reinforcement learning: theoretical framework and an algorithm. Paper presented at the Proceedings of the 15th International Conference on Machine Learning, Madison Wisconsin, 24–27 July 1998
17. Greenwald A, Hall K (2003) correlated-q learning. Paper presented at the Proceedings of the 20th International Conference on Machine Learning, Washington DC, 21–24 Aug 2003
18. Gao D (1998) On structures of supervised linear basis function feedforward three-layered neural networks. Chin J Comput 21(1):80–86

Chapter 6
TD-LTE Based CWN Testbed

Although cognitive radio technologies have been proposed and studied for many years, the implementation of cognitive radio technologies in cellular networks still faces many challenges. To guarantee the continuous services in cellular networks, traditional cognitive radio technologies which use the silent period scheme for spectrum sensing while stop the communication services are no longer effective and applicable considering stringent delay constraints for voice and real-time video services. Furthermore, in terms of the dynamic changing feature of vacant spectrum resources, the fast and low complexity vacant spectrum awareness technologies, such as the spectrum sensing and geo-location database, are prerequisites for the implementation of CWNs. Moreover, the rapid spectrum handover and reconfiguration abilities are the key technologies to support the dynamic and efficient spectrum utilization based on the vacant spectrum awareness results. Therefore, new technologies are designed and deployed for the TD-LTE based CWNs. To obtain the global radio environment information, the fast and accurate spectrum information awareness technology should be designed, which can improve the efficient dynamic spectrum access and utilization in CWNs. Furthermore, to guarantee the QoS of users in TD-LTE based CWNs with specific delay constraints, the guard period in TD-LTE system is utilized for spectrum sensing without the silent period and communication service stop. And seamless spectrum handover and dynamic bandwidth adjustment technologies are also proposed by considering both the packet loss and service drop rate. Therefore, both challenges and solutions on how to realize the TD-LTE based CWNs are analyzed with simulations and field results in this chapter.

6.1 Scenario of TD-LTE Based CWN Testbed

As shown in Fig. 6.1, two cells of TD-LTE based CWN are deployed and each cell includes one cognitive eNB (CeNB) and four CR users (UEs). The Advanced Spectrum Management (ASM) server obtains the available spectrum resources information from the Geo-location database which stores and updates the latest

© The Author(s) 2015
Z. Feng et al., *Cognitive Wireless Networks,* SpringerBriefs in Electrical and Computer Engineering, DOI 10.1007/978-3-319-15768-9_6

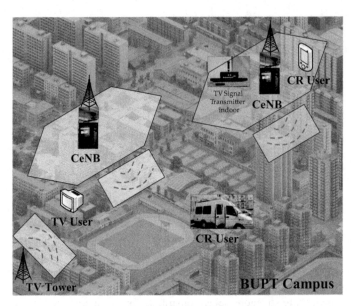

Fig. 6.1 Scenario of TD-LTE based CWN testbed in BUPT

vacant spectrum information from local spectrum sensing results of CeNBs. Then, the ASM server is responsible for vacant spectrum reallocation and coordination among different CeNBs in a long period of time.

Based on the proposed strategies for the dynamic spectrum management in CWN, the spectrum allocation and management are performed in two steps. First, the ASM is responsible for the long-term spectrum decision making, which assigns spectrum bands to each CeNB in a long period of time. Second, the spectrum decision of the ASM is based on the spectrum regularity policies and knowledge from the geo-location database. Finally, the efficiency of spectrum utilization can be improved by about 30 % using the dynamic spectrum sharing scheme among CeNBs under the control of ASM server in the laboratory scenario [1].

The description of the scenario of TD-LTE based CWN testbed is shown in Fig. 6.2 [1]. And the TD-LTE based CWN testbed mainly consists of two levels, the lower level equipments and the higher level equipments. The formers of the TD-LTE based CWNs and the signal generator system of the primary user are responsible for the fundamental wireless communication and the spectrum sensing. The latters, including the ASM and the geo-location database, take charge of the dynamic spectrum allocation and the management of the whole spectrum band.

First, the TD-LTE based CWNs in lower level consist of two cognitive cells, each of which includes one cognitive eNB (CeNB) and four CR users (UEs). Due to the limitation of the UE's power consumption, the UE only performs the spectrum handoff after receiving the message from the CeNB. The CeNB is responsible for the real-time spectrum sensing of the whole selected spectrum band as well as the normal wireless schedule and the transmission control function.

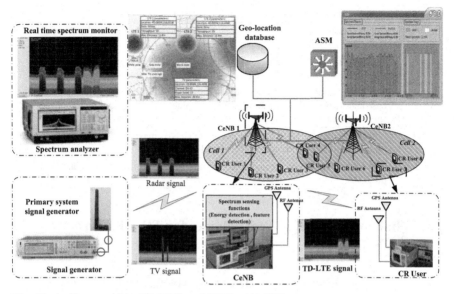

Fig. 6.2 Architecture of the TD-LTE based CWN testbed [1]

Second, the primary system signal generator cannot be constrained by the TD-LTE based CWNs in design and the secondary user must not influence the QoS of the primary system communication. The TD-LTE based CWNs testbed uses the software controlled signal generator equipment to generate and transmit TV signals to rebuild the real radio environment.

Third, the geo-location database collects and stores the policy information and the real radio environment information in the whole spectrum band at different locations by the previous simulation and measurement work. On one hand, the ASM gets the available spectrum band information from the geo-location database. On the other hand, the geo-location database also receives and updates the latest spectrum band information from the local spectrum sensing results of CeNBs.

Finally, the ASM server is responsible for vacant spectrum reallocation and co-ordination among CeNBs in long term. In the CWN testbed, the proposed strategies of the dynamic spectrum management performs the spectrum allocation function in two steps. The ASM is responsible for the long-term spectrum decision which assigns spectrum bands to each CeNB in a long period of time. And the spectrum decision made by the ASM is based on the spectrum regularity policies and knowledge learning from the geo-location database. In general, if the traffic in one CeNB is so heavy that there are not enough available spectrum resources, the long-term spectrum decision will assign extra spectrum which is formerly assigned to other CeNB or the vacant spectrum in the permitted spectrum band. On the contrast, the CeNB with low traffic can deliver the same services in a good quality to the end users and provide redundant spectrum to other CeNB. The working procedure of the proposed TD-LTE based CWN testbed is depicted in Table 6.1 below.

Table 6.1 Major procedure of the TD-LTE based CWN testbed

Step 1:	CeNBs perform the spectrum sensing in the selected spectrum band of the CWN testbed
Step 2:	When the spectrum utilization situation changes, CeNBs report the new occupation of the spectrum band to the Geo-location database
Step 3:	The ASM periodically gets the relative information from the Geo-location database and helps CeNBs to choose another available working band roughly in a wide spectrum range
Step 4:	After receiving the ASM decision message, CeNBs make a more detail and accurate choice on the working spectrum band
Step 5:	CeNBs perform the spectrum handover and report the final working condition to the ASM

6.2   Challenges and Key Technologies for TD-LTE Based CWN Testbed Implementation

Considering the implementation of TD-LTE based CWN testbed, challenges still need to be overcome, such as the protocol modifications, fast and accurate spectrum sensing, rapid spectrum reconfiguration for the seamless spectrum handover and efficient utilization of the scattered spectrum resources. In order to overcome these challenges, solutions with key technologies are proposed in this chapter, including the protocol modification and stack design, radio environment information cognition, seamless spectrum handover, dynamic bandwidth adjustment, and efficient spectrum utilization technologies as described in detail below.

6.2.1   Protocol Modification in TD-LTE Based CWN Testbed

The first challenge is how to make modifications to protocol and stack in order to support the cognitive radio technologies in TD-LTE system. In order to overcome these challenges, we designed the Cognitive Channel and modified the Physical Layer Protocol based on the 3rd Generation Partnership Project (3GPP) TS 36 serial technical specifications R9 [2–4] in the single input and single output mode.

6.2.1.1   Protocol Modification and Stack Design

The goal of the CWN testbed is mainly to introduce the cognitive loop and prove the high efficiency of the dynamic spectrum usage. So how to introduce the cognitive cycle into the TD-LTE system with minimal changes to the existing LTE protocol stack remains to be a challenge. This section describes the detail adjustments based on the original LTE protocol stack.

The TD-LTE based CWNs is built based on the 3GPP TS 36 serial technical specifications R9 [2–4] in the single input single output mode. In the realization of the CWN testbed, the cognitive message in the control signal of dynamic spectrum access is added in the existing communication protocols to simplify the 3GPP TS 36 serial technical specifications [2–4] to reduce the complexity of the whole TD-LTE cellular system.

6.2.1.2 Cognitive Channel Design

As the 3GPP LTE protocol stack sets, the transmission process of the control message from CeNBs to UEs passes through the corresponding logic channel, transport channel and physical channel before the final signal emitting. So the CWN testbed adds the cognitive channels to send the cognitive message according to the transmit rules of the 3GPP TS 36 serial technical specifications [2–4]. Figure 6.3 shows the realization of the cognitive information delivery channels.

1. Cognitive Control Channel (CogCCH)
   The CogCCH is added to the logic control channel as same as the Broadcast Control Channel (BCCH), Dedicated Control Channel (DCCH) and Dedicated Traffic Channel (DTCH). And only the transparent transmission mode (TM) is supported.
2. Cognitive Channel (CogCH)
   The CogCH is added to the transport channel as same as the Broadcast Channel (BCH) and Shared Channel (SCH). And the HARQ function is not supported in the Medium Access Control (MAC) layer.
3. Physical Cognitive Channel (PCogCH)
   The PCogCH is added to the physical channel as the working method of the Physical Broadcast Channel (PBCH). All UEs in the CWN need to demodulate the PCogCH in order to avoid the consumption of the same data transporting in each UE's Physical Shared Channel (PSCH).

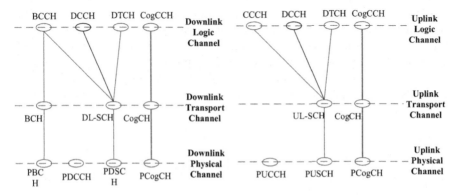

Fig. 6.3 Cognitive channel on downlink and uplink

6.2.1.3   Physical Layer Protocol Modification

In the physical layer, the TD-LTE based CWNs apply the fixed frame structures and make necessary simplifications accordingly. In Fig. 6.3, the PBCH, PDSCH, PDCCH in the downlink and the PUSCH, PUCCH in the uplink are used and the code and modulation procedures are explained in detail.

1. PBCH

   Main Information Block (MIB) is the message transmitted on the PBCH. Figure 6.4 shows the MIB's process and we abandon the PHICH bits and expand the bandwidth bits to 10 bits to avoid the blind detection of the PBCH.

2. PDCCH

   The TD-LTE based CWNs only keep the two necessary Downlink Control Information (DCI) format, including the format 0 and format 1. As each TD-LTE cell only supports four UE, the PDCCH distributes the fixed mapping location for each UE to avoid the blind detection.

3. PDSCH

   The processing procedures of PDSCH are the same as the 3GPP sets. Because of the added PCogCCH, we should redesign the mapping scheme of the sequence to resource elements (RE). Figure 6.5 shows the RE mapping when the bandwidth is set to 20 MHz. We choose the 4 MHz band in the middle as the authorized spectrum band in the consideration of balancing the fewer symbols the PDCCH occupies and the more payload of the PDSCH. Moreover, our mapping scheme also follows the mapping of the Primary Synchronous Signal, the Secondary Synchronous Signal and the PBCH in the existing protocols.

6.2.2   Radio Environment Information Cognition

To verify the performance of key cognitive radio functions, fundamental challenges need to be solved, including the capability of working in a wide dynamic radio frequency range to support different operation frequencies, fast and accurate spectrum sensing, and rapid spectrum reconfiguration for the spectrum handover. In order to overcome these challenges, a novel spectrum sensing module is proposed. In contrast to the silent period scheme in traditional cognitive radio system, a novel radio environment information cognition scheme is proposed without using the silent period. Instead, the guard period between the uplink and downlink data transmission period is used for spectrum sensing without extra delay to guarantee users' QoS.

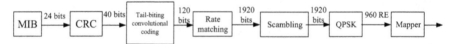

Fig. 6.4 Transport block processing for PBCH

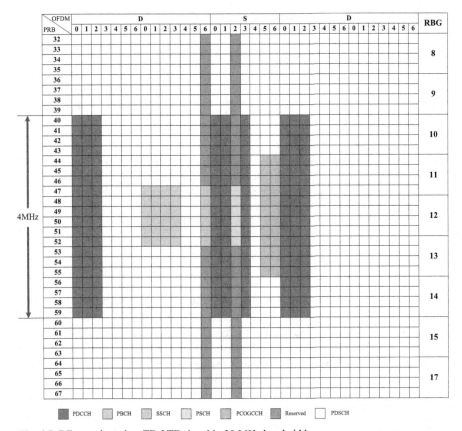

Fig. 6.5 RE mapping when TD-LTE signal in 20 MHz bandwidth

Moreover, the feature detection scheme is proposed for the TV signal and radar signal, which can decrease the computation complexity and time consumption with more than 99 % accuracy for the real-time spectrum sensing in the TD-LTE based CWN testbed. Moreover, the multi-domain environment information, such as the wireless domain, network domain, user domain and policy domain, is obtained by the CeNB from the results of spectrum sensing and the database of operators, which is stored in the multi-domain cognitive database in the TD-LTE based CWN testbed.

First, in the CWN testbed, the spectrum sensing technology in CeNBs is utilized not only to find the spectrum bands that are not being used by the primary user at a particular time in a particular geographic area, but also to detect the burst of the primary user's signal as soon as possible in order to avoid the interference from TD-LTE system. Compared to other spectrum sensing technologies, the energy detection and the feature detection have less computing complexity and less time consumption while functioning accurately with more than 99 % accuracy of the real-time spectrum sensing in the CWN testbed. The feature detection is applied in 702–798 MHz band because of the distinct feature of the analog TV signal.

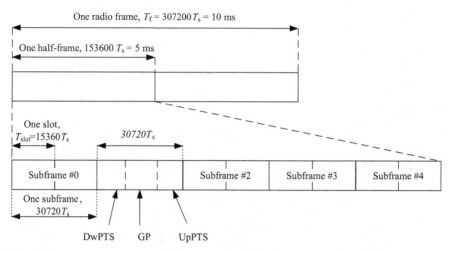

Fig. 6.6 Frame structure in the TD-LTE based CWN [2]

Figure 6.6 shows the TD-LTE frame structure, where one radio frame consists of the subframe "D" for downlink transmissions, the subframe "U" for uplink transmissions and a special subframe "S". On the one hand, it will certainly lead to more than 10% waste of the payload in the transmissions if either the subframe "D" or "U" is utilized. So we insert the spectrum sensing time in the special subframe "S", which includes three fields, i.e., Downlink Pilot Time Slot (DwPTS), Guard Period (GP) and Uplink Pilot Time Slot (UpPTS). On the other hand, the GP in the subframe "S" is used for insuring the uplink synchronization of all UEs and all equipments do not emit the signal in the GP. The GP decides the coverage of the CeNB and the UEs are different in their GP time as the different distance from the UE to the CeNB. Therefore, the UEs can not be utilized to complete the spectrum sensing due to the mutual influence.Therefore, we add the spectrum sensing function only in the TD-LTE CeNB at the time of the CeNB's GP which means the longer GP, the longer time the CeNB can use to fulfill the spectrum sensing. Considering the time consumption of realizing the spectrum sensing algorithm in DSP and reporting the spectrum sensing result back to the master controller, the spectrum sensing time in this TD-LTE based CWN testbed is 2 ms in every 10 ms. The proposed value of the sensing cycle from FCC [5] is 30 s in idle time or 60 s in processing time and there is no requirement of the spectrum sensing time from FCC.

The spectrum sensing module of the CWN testbed uses an engineering method to set up the threshold between the primary user's signal and the noise. First, we set the transmission power of the primary user's sender to the maximum value and ensure the reliable communication of the primary user's signal on one channel. Second, we increase the distance between the primary user's signal sender and receiver and reduce the transmission power. As the increasing of the distance and the reducing of the radio power, the quality of the primary user's communication is ensured above a certain degree until the primary user's signal receiver cannot

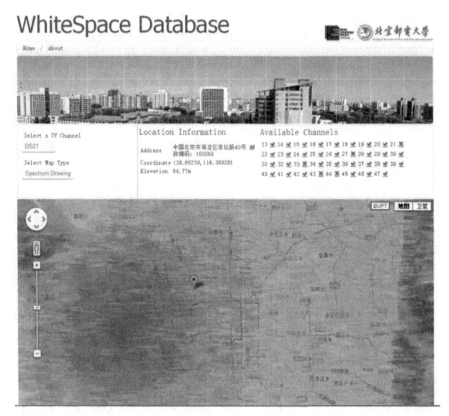

Fig. 6.7 Multi-domain database on website

accurately distinguish the signal from the sender and the noise. Then, we set the average value of the ratio between the carriers' energy and the whole received signals' energy when the communication quality declines to a certain degree on all different channels as the spectrum sensing threshold in the whole spectrum band. In this TD-LTE based CWN testbed, the accuracy of the spectrum sensing threshold is − 120 dBm, while the proposed value from FCC [5] is − 114 dBm.

Second, the multi-domain database will collect the information in wireless domain, network domain, user domain and policy domain using the TD-LTE based CWN testbed. In terms of the dynamic changing feature of wireless domain information, the implementation software system is proposed and depicted in Fig. 6.7. Considering the TV system from 700 to 800 MHz in the urban area of Beijing city, both the occupied and vacant spectrum resources are depicted in accordance to different locations. First, we obtain the number, location, transmission power and channel parameters of the TV tower in Beijing. Then, based on the real geographical features of Beijing, we simulate the radio environment of the TV signals on each channel and correct the mathematical model by the real measurement results in the urban area. Third, the multi-domain database includes all the simulation values of TV signals on all channels and the policy information of the spectrum band in Beijing. Figure 6.7 shows the primary signal information from the visualization interface of the multi-domain database on the web browser. The multi-domain database responses the information of the TV signal on Channel 21 in the urban area of Beijing where different colors depict different received power values of TV signals. This database can also show the available channels on different locations using different checkboxes at the top right part in Fig. 6.7.

6.2.3   Seamless Spectrum Handover

Considering the scattered vacant spectrum resources and the random appearance of primary users' activity, the TD-LTE based CWN testbed should be able to achieve the seamless spectrum handover among different frequency bands to efficiently utilize the spectrum resources. As a major challenge, the time delay during the spectrum handover process is a key performance indicator which will affect the performance of services. Therefore, a novel seamless spectrum handover technology is proposed which can achieve the handover time less than 50 ms for TD-LTE based CWN testbed, in contrast to the proposed time delay parameter of 2 s in FCC [5].

When the CeNB detects that the spectrum band currently used by the TD-LTE system has been occupied by a primary user, both CeNB and UEs will change the operating spectrum band after CeNB receives the spectrum handover message from the ASM. First, the CeNB should choose an appropriate center frequency and bandwidth for the TD-LTE system because the messages from the ASM only provide available spectrum bands in the long term. Then, as the UE do not have the spectrum sensing function, the CeNB should inform the UEs the new center frequency, the bandwidth and the spectrum handover time in the spectrum handover message through the cognitive channel before the spectrum handover. After UEs receive the spectrum handover messages, both CeNB and UE in one cell perform spectrum handover at the same time and the communication of the TD-LTE system continues on the new spectrum band. But the signals of the TD-LTE system and the primary user will interfere each other when the primary user's signals emerges, so the transmission of the spectrum handover messages in the TD-LTE system becomes unreliable. If the UE cannot receive the spectrum handover message in time, it will greatly increase the time cost of the spectrum handover and consequently interfere the communication of the primary users seriously. In a worse condition that the UE does not change the operating spectrum band in accordance with the CeNB in the same cell, the whole TD-LTE system will not work anymore. In order to overcome this problem, the UEs in the CWN testbed will sweep the whole spectrum band such as the whole 702–798 MHz in the TV scenario when the transmission of the TD-LTE system is failed. The step size of the sweeping process is set to 4 MHz in the TV scenario because the center frequency point is only set on the center and the edge of each channel.

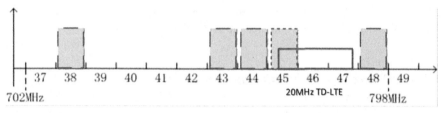

Fig. 6.8 Spectrum handover scenario

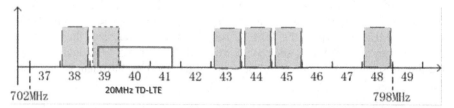

Fig. 6.9 Spectrum handover scenario

When CeNB detect that the TV system begins a new transmission on the spectrum band of the current TD-LTE system, the TD-LTE based CWNs will change the operating RF band to another three continuous vacant channels in order to avoid the interference to the transmission of the TV system. The result of spectrum handover is shown in Figs. 6.8 and 6.9, where the block in white stands for the TD-LTE signal and the block in grey stands for the TV signal.

1. The TD-LTE signal with a 20 MHz bandwidth works on the Channel 45, Channel 46 and Channel 47. And the TV signal works on the Channel 38, Channel 43, Channel 44 and Channel 48.
2. The TV signal appears on the Channel 45.
3. The TD-LTE signal with a 20 MHz bandwidth changes to work on the Channel 39, Channel 40 and Channel 41.
4. The TV signal appears on the Channel 39.
5. The TD-LTE signal with a 20 MHz bandwidth changes to work on the Channel 40, Channel 41 and Channel 42.

The time of the spectrum handover depicts the performance of the TD-LTE based CWN testbed. The smaller value of the time cost in the whole spectrum handover process means less interference to the primary user's communication system.

The time cost of the spectrum handover is influenced by many factors in the whole process, including the delay of the transmission of the spectrum handover message, the time cost of receiving the spectrum decision from the ASM and the communication time between the RF modules, the baseband unit and the master controller. Considering the total time cost, we set the time of the spectrum handover from CeNB to UEs by 50 ms after the CeNB has detected the appearance of the primary user's signal. In contrast to the proposed time of the handover time from FCC

[5] of 2 s, the handover time of 50 ms for the proposed TD-LTE based CWN testbed is much smaller, which can guarantee the service quality of users.

6.2.4 Dynamic Bandwidth Adjustment

In terms of the scattered and discontinuous vacant spectrum resources, the dynamic bandwidth adjustment technology is applied by TD-LTE based CWN testbed in order to efficiently utilize the vacant spectrum resources. When CeNBs detect that there are not enough vacant spectrum bands to support the current TD-LTE system with a 20 MHz bandwidth, the TD-LTE based CWNs will decrease the operating bandwidth to a smaller one, such as 15 MHz, 10 MHz and 5 MHz and continue the transmission for the end users. The result of bandwidth adjustment is shown in Figs. 6.10 and 6.11.

1. The TD-LTE signal with a 20 MHz bandwidth works on the Channel 40, Channel 41 and Channel 42. And the TV signal works on the Channel 38, Channel 39, Channel 43, Channel 44, Channel 45 and Channel 48.
2. The TV signal appears on the Channel 41.
3. Because there are no three continuous vacant TV channels, so the TD-LTE signal changes to 15 MHz bandwidth and works on the two continuous channels of Channel 46 and Channel 47.
4. The TV signal appears on the Channel 47.
5. Because there are no two continuous channels, the TD-LTE signal changes to 5 MHz bandwidth and works on the Channel 37.

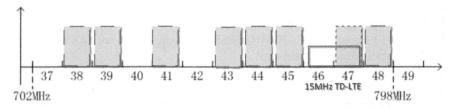

Fig. 6.10 Bandwidth adjustment scenario

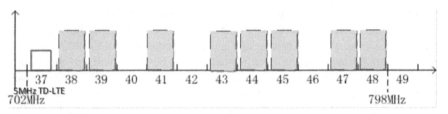

Fig. 6.11 Bandwidth adjustment scenario

Therefore, by using the dynamic bandwidth adjustment scheme, the proposed TD-LTE based CWN testbed can efficiently utilize the scattered spectrum resources by adjusting the system bandwidth in accordance to the spectrum vacancy status on TV white spaces.

6.2.5   Efficient Spectrum Utilization

To improve the efficiency of spectrum utilization in CWN, both the dynamic spectrum allocation and multi-flow transmission technologies are proposed and described in detail below.

6.2.5.1   Dynamic Spectrum Allocation and Efficient Spectrum Utilization Technology

By using the TD-LTE based CWN testbed, the vacant spectrum resource utilization can be improved efficiently by using the ASM function. As shown in Fig. 6.1, two cells are deployed and each cell includes one CeNB and four CR UEs. The ASM server obtains the available spectrum resources information from the geo-location database which stores and updates the latest vacant spectrum information from local spectrum sensing results of CeNBs. Then, the ASM server is responsible for vacant spectrum reallocation and coordination among different CeNBs in a long period of time. Based on the proposed dynamic spectrum management technology in CWN, the spectrum allocation and management are performed in two steps. First, the ASM is responsible for the long-term spectrum decision which assigns spectrum bands to each CeNB in a long period of time. Second, the spectrum decision made by the ASM is based on the spectrum regularity policies and knowledge learning from the geo-location database. Finally, the efficiency of spectrum utilization can be improved by about 30 % using the dynamic spectrum sharing scheme among CeNBs under the control of ASM server in the laboratory scenario in [1].

6.2.5.2   Multi-flow Transmission Technology

By implementing the multi-flow transmission technology, the service experience of users with high-bandwidth demand and the resource utilization of wireless network can be improved dramatically.

Figure 6.12 shows a typical feedback-based LTE-WLAN multi-flow transmission scenario. Multi-flow transmission server and multi-mode terminals are the nodes that have the capability of supporting multiple communication modes, which enable the data transmission via WLAN AP and LTE base station nodes simultaneously.

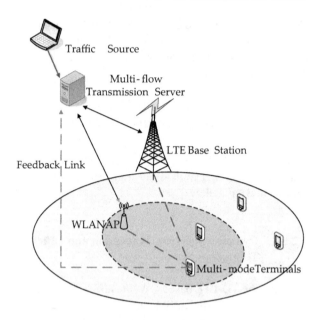

Fig. 6.12 Feedback-based multi-flow transmission scenario

The basic procedure of multi-flow cooperative transmission can be described in three steps. First, initial data packets from the traffic source are split into multiple packets by specific multi-flow algorithm from the transmission server. Second, these data packets with various destination networks are transmitted to terminals through multiple heterogeneous wireless networks. Finally, multi-flow data aggregation is implemented in the multi-mode terminal.

Based on existing strategies, the multi-flow transmission system can be divided into network-side and terminal-side operations. In the terminal-side, according to the quality of received traffics, multi-mode terminals report information about real-time transmission state of all accessible networks, such as latency, packet loss, to the multi-flow transmission server through a feedback link. In the network-side, based on the feedback information, the multi-flow algorithm is executed by multi-flow transmission server to make reasonable multi-flow transmission strategy about network selection and the intensity of multi-flow. The quality of service can be improved greatly by integrating the capacity among independent heterogeneous networks.

6.3   Concluding Remarks

The TD-LTE based CWN testbed successfully introduces the cognition cycle into the TD-LTE system, which verifies the feasibility of the coexistence of the TD-LTE based CWN and the primary communication system. To verify the theoretical research works, this chapter describes the procedures of designing and developing the TD-LTE based CWN testbed which can improve the spectrum utilization. The architecture of the CWN testbed is introduced and the simplified protocol stack design is proposed thereafter. Finally, key technologies such as spectrum sensing, network information collection, spectrum handover using the TD-LTE based CWN testbed are analyzed thoroughly.

References

1. Zhang P, Liu Y, Feng Z Y et al (2012) Intelligent and efficient development of wireless networks: a review of cognitive radio networks. Chin Sci Bull 57(28–29):3662–3676
2. 3GPP, 3rd Generation Partnership Project (2011) Technical Specification Group Radio Access Network; Evolved Universal Terrestrial Radio Access (E-UTRA); Physical Channels and Modulation (Release 9), TS 36.211
3. 3GPP, 3rd Generation Partnership Project (2011) Technical Specification Group Radio Access Network; Evolved Universal Terrestrial Radio Access (E-UTRA); Multiplexing and channel coding (Release 9), TS 36.212
4. 3GPP, 3rd Generation Partnership Project (2011) Technical Specification Group Radio Access Network; Evolved Universal Terrestrial Radio Access (E-UTRA); Physical layer procedures (Release 9), TS 36.213
5. Federal Communications Commission (2010) Unlicensed operations in the TV broadcast bands. 2nd memorandum opinion and order, ERR docket NO. 10–174

Chapter 7
Standardization Progress

The current radio environment is characterized by its heterogeneity. Different aspects of this heterogeneity include: multiple operators and services, various radio access technologies, different network topologies, broad range of radio equipments, and multiple spectrum bands. Therefore, both technical challenges and business opportunities exist and show a broad way for further study. To exploit such opportunities, the concept of Cognitive Radio System (CRS) has been developed with different use cases and business models. This has triggered the progress in standardization activities at all levels, including ITU, IEEE, and ETSI. All of these organizations consider multiple CRS deployment scenarios and business directions.

7.1 ITU

Before World Radio Conference (WRC-07), a common concern of ITU-R was the protection of existing services from potential interference from software defined radio (SDR) and CRS systems. WRC-07 adopted new agenda item 1.19 requiring WRC-12 to consider regulatory measures for the introduction of SDR and CRS. Moreover, the WRC-07 adopted Resolution 956 (WRC-07), inviting ITU-R to study the need for regulatory measures related to the application of SDR and CRS. The agenda item 1.19 is "to consider regulatory measures and their relevance, in order to enable the introduction of software-defined radio and cognitive radio systems, based on the results of ITU-R studies, in accordance with Resolution 956 (WRC 07)." In this regard, WRC-07 adopted Resolution 956 (WRC-07) on "Regulatory measures and their relevance to enable the introduction of software-defined radio and cognitive radio systems". The resolution also resolved that WRC-12 should consider the results of the studies and take appropriate actions. Therefore, in 2012, both the Radiocommunication Assembly (RA-12) and World Radio Conference (WRC-12) are held by ITU and the studies on the implementation and use of cognitive radio systems are achieved in the resolution ITU-R 58 (RA-12) [1]. In terms of the flexibility and improved efficiency of the overall spectrum utilization, both the advantages and potential coexistence with existing systems among radiocommunication

© The Author(s) 2015
Z. Feng et al., *Cognitive Wireless Networks,* SpringerBriefs in Electrical
and Computer Engineering, DOI 10.1007/978-3-319-15768-9_7

services should be taken into account with the introduction of CRS technology, such as the frequency band, technical and operational characteristics. Moreover, the recommendation 76 (WRC-12) on the deployment and use of cognitive radio systems is achieved in [2], which considers about the implementation and use of CRS in accordance with the resolution ITU-R 58 in [1]. And the radio system implementing CRS technology needs to operate in accordance with the provisions of the Radio Regulations (RR) in [2]. The task of conducting necessary studies is assigned to ITU-R WP 1B. Also, ITU-R WP 5A and 5D study the technical aspects of Cognitive Radio Systems [3].

7.1.1   ITU WP 1B

Within agenda item 1.19, ITU-R WP1B has developed two documents related to CRS:
Definitions of SDR and CRS [4].
Draft CPM text on WRC-12 agenda item 1.19 [5].

7.1.1.1   Definitions of SDR and CRS

Under the framework of WRC-12 agenda item 1.19, ITU-R WP1B has developed definitions of SDR and CRS to assist in the conduct of studies, to make clear the distinctions between SDR and CRS technologies, and to provide a common understanding and facilitate their use in an unambiguous way in ongoing work by the ITU-R. Below are these two definitions.

Software-defined Radio (SDR) "A radio transmitter and/or receiver employing a technology that allows the RF operating parameters including, but not limited to, frequency range, modulation type, or output power to be set or altered by software, excluding changes to operating parameters which occur during the normal pre-installed and predetermined operation of a radio according to a system specification or standard."

Cognitive Radio System (CRS) "A radio system employing technology that allows the system to obtain knowledge of its operational and geographical environment, established policies and its internal state; to dynamically and autonomously adjust its operational parameters and protocols according to its obtained knowledge in order to achieve predefined objectives; and to learn from the results obtained."

7.1.1.2   Draft CPM Text on WRC-12 Agenda Item 1.19

With regard to CRS, the following key elements can be identified in the draft CPM text.

1. CRS deployment scenarios
   ITU-R WP1B has identified four deployment scenarios:

   1) Use of CRS technology to guide the reconfiguration of connections between terminals and multiple radio systems
   2) Use of CRS technology by an operator of radio communication systems to improve the management of its assigned spectrum resources
   3) Use of CRS technology as an enabler of cooperative spectrum access
   4) Use of CRS technology as an enabler of opportunistic spectrum access

2. CRS challenges and opportunities
   A common concern within the ITU-R is the protection of existing services from potential interference from the services implementing CRS technology, especially from the dynamic spectrum access capability of CRS.
   In addition, a service using CRS should not adversely affect other services in the same band with the same or higher status. Thus, the introduction and operation of stations using CRS technologies should not impose any additional constraints to other services sharing the band.

3. CRS capabilities and applicabilities to facilitate coexistence in shared bands
   The ITU-R has identified a set of capabilities of CRS that may facilitate coexistence with existing systems. The following elements could be considered as examples of capabilities of CRS:
   - spectrum sensing
   - positioning (geo-location)
   - access to information on the spectrum usage, local regulatory requirements and policies
   - capabilities to adjust operational parameters

4. Relationship between SDR and CRS
   SDR is recognized as an enabling technology for the CRS. SDR does not require characteristics of CRS for operation. Either technology can be deployed/implemented without the other. In addition, SDR and CRS are at different phases of development, i.e. radio communication systems using applications of SDR have been already utilized and CRS is now being researched and applications are under study and trial.

5. Analysis of the results of studies
   The potential benefits and applicabilities of CRS technologies to various radio communication services are recognized as well as the fact that CRS would be introduced in some services.
   There is a need for further studies on CRS technology, addressing dynamic and/or opportunistic spectrum access.
   The use of CRS in some bands by particular radio communication services may require the development of unique ITU-R Recommendations and Reports to address these issues. However, the study concluded that there is no need for modification to the Radio Regulations for this Agenda item for the introduction of CRS technology.

6. Methods to satisfy the agenda item Two methods are proposed to satisfy the
   agenda item:

- No change to the Radio Regulations
- Add a WRC Resolution providing guidance for further studies and guidance for
  the use of CRS and no other changes to the Radio Regulations.

7.1.2   ITU WP 5A

Within agenda item 1.19, ITU-R WP5A has developed a document related to CRS:
Introduction to cognitive radio systems (CRS) in the land mobile service (LMS) [6].
Apart from that, working document towards a preliminary draft new report ITU-R
LMS.CRS2 is under development.

The LMS.CRS Report addresses the cognitive radio systems in the LMS above
30 MHz, excluding international mobile telecommunications (IMT). It provides a
general description of CRS addressing technical features and capabilities, potential
benefits and technical challenges. It also describes a set of deployment scenarios.

CRS is not a radio communication service, but rather a system that employs
technology that may be implemented in a wide range of applications in the LMS.
A system in the LMS using CRS technology must operate in accordance with the
Radio Regulations governing the use of a particular band, as is also the case for any
system in the land mobile or in any other service.

The report also addresses the potential benefits of CRS technology including
a possible improvement in the efficiency of spectrum utilization and facilitation
of additional flexibility. The implementation of CRS technology may introduce
challenges of technical or operational nature. These challenges could include implementation complexity, reliability of different methods for obtaining knowledge and
interference avoidance.

Besides, this LMS.CRS Report addresses four CRS deployment scenarios. Their
implementation and feasibility will depend upon the resolution of technical challenges and compliance with national and ITU Radio Regulations:

1. Use of CRS technology to guide reconfiguration of connections between
   terminals and multiple radio systems.
2. Use of CRS technology by an operator of a radio communication system to
   improve the management of its assigned spectrum resource.
3. Use of CRS technology as an enabler of cooperative spectrum access.
4. Use of CRS technology as an enabler for opportunistic spectrum access in bands
   shared with other systems and services.

7.1.3 ITU WP 5D

Within agenda item 1.19, ITU-R WP5D has developed a document related to CRS: Cognitive radio systems specific for International Mobile Telecommunications systems [7].

This document addresses aspects of cognitive radio systems specific to IMT systems. It includes results of studies to determine the impact of adding cognitive radio capabilities to existing IMT systems, and analyses the benefits, challenges and impacts of CRSs in IMT, including a description of how the systems would be used in IMT system deployments and their possible impacts on the use of IMT spectrum.

The implementation of CRS technology in IMT networks will progress stepwise due to a number of technical challenges coupled with the current state of the technology. In addition, the implementation of CRS technology in IMT networks may introduce specific and unique challenges of technical or operational nature.

The challenges related to the introduction of CRS technology inside IMT systems are summarized as follows.

- Interference management: so the existing radio systems users do not suffer harmful interference due to any CRS operation;
- QoS: the current QoS level of existing radio systems users should be guaranteed in case of any CRS operation;
- Reliability: the CRS operations should be entirely reliable for the users and all the involved nodes in the system(s);
- Mobility: a full seamless connection experience should still be guaranteed to users in case of any CRS operation;
- Timing: the CRS operations should be executed and signaled to all the affected users and nodes in a timely manner;
- Security: the existing radio systems should have a sufficient degree of protection against malicious behavior which may arise due to any CRS operation and in particular to guarantee that user devices will not bypass network policies.

The report also addresses the issues related to the potential benefits, and key performance indicators of using CRS technology in IMT systems.

1. Potential benefits of using CRS technology in IMT systems

    1. Overall spectrum utilization and capacity improvement
    2. Radio resources utilization flexibility
    3. Interference mitigation

2. Potential implications of using CRS technology in IMT systems

    1. Signaling overhead in CRS
    2. Increase of the system complexity
    3. Increase of the control/user plane latency

3. Key performance indicators for CRS technology in IMT systems

In order to evaluate the performance of different methods, the unique key performance indicators for spectrum knowledge, overall system performance, decision making, and user experience of CRS technology in IMT systems may need to be defined. With specific reference to overall system performance aspect, the advantages and implications brought to IMT systems by CRS technology can be uniquely expressed as some parameters, such as channel set up and release duration, signaling load, cell average throughput, overall network capacity and reliability.

7.2   IEEE Standards

7.2.1   IEEE 802.22

One of the first IEEE Working Groups (WGs) to consider CR technology was IEEE 802.22, created in 2004 and developing a standard for wireless regional area networks (WRANs) using white spaces in the TV spectrum [8].

The IEEE 802.22 WRAN is an infrastructure cellular network where the BS covers an area of radius spanning from 30 km (typical) to 100 km. The WRAN end user is referred to as Customer Promise Equipment (CPE) whose transceivers are installed on a house. A conceptual illustration of IEEE 802.22 is provided in Fig. 7.1 [8].

The WRAN is designed to provide throughputs of 1.5 Mb/s in the downstream and 384 kb/s in the upstream, and its PHY utilizes OFDM modulation to overcome possibly excessive delays in a wide coverage area. In addition, it provides PU protection such as spectrum sensing and a geo-location database for PU-SU coexistence, and also supports self-coexistence between WRANs via the Coexistence Beacon Protocol (CBP) [8].

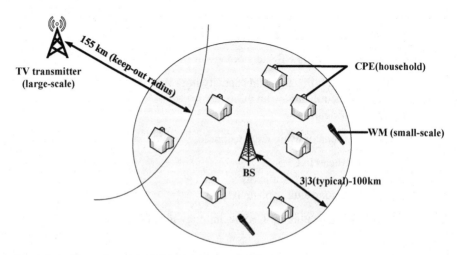

Fig. 7.1 An illustration of the IEEE 802.22 network [8]

7.2.2 IEEE 802.11af

In 2008 Google and Microsoft announced their interests in using TV white spaces (TVWS) for an enhanced type of Wi-Fi like Internet access. The idea was later formalized as a new standard called IEEE 802.11af, for which an 802.11 task group was chartered. 802.11af is expected to provide much higher speed and wider coverage than current Wi-Fi, thanks to the better propagation characteristics of the VHF/UHF bands [8].

IEEE 802.11af can be modeled as a wireless network with a CR-enabled access point (AP) and associated CR devices as terminals. The CR APs operate on WS via spectrum trading schemes, and the incurred time varying spectrum availability introduces new challenges. For example, upon appearance of PUs in a leased channel, the AP should relocate the CRs in the channel, which requires eviction control of in-service customers in case the remaining idle channels cannot accommodate all the spectrum demands [8].

Although Wi-Fi over WS is still in its infancy, its resemblance to today's Wi-Fi hotspots suggests that it may become a "killer application" in CR-based wireless networks. By utilizing more favorable spectrum bands than the ISM, the new Wi-Fi must be able to support QoS guarantees and resource-intensive multimedia services more easily than the current Wi-Fi [8].

7.2.3 IEEE DySPAN-SC 1900

The IEEE P1900 Standards Committee, DySPAN-SC's predecessor, was established in the first quarter 2005 jointly by the IEEE Communications Society (ComSoc) and the IEEE Electromagnetic Compatibility (EMC) Society. The objective of the effort was to develop supporting standards dealing with new technologies and techniques being developed for the next generation radio and advanced spectrum management.

On 22 March 2007, the IEEE Standards Association Standards Board approved the reorganization of the IEEE 1900 effort as Standards Coordinating Committee 41 (SCC41) "Dynamic Spectrum Access Networks (DySPAN)". The IEEE Communications Society and EMC Society were supporting societies for this.

IEEE SCC41 was approached by the IEEE ComSoc Standards Board (CSSB) in late 2010, as ComSoc Standards Board was extremely interested in SCC41 being brought back directly under its wing. SCC41 voted to be directly answerable to ComSoc in December 2010, and was thereby renamed as IEEE DySPAN-SC. At its December 2010 Meeting, the IEEE Standards Association Standards Board (SASB) approved the transfer of projects from SCC41 to CSSB.

The objective of IEEE DySPAN-SC is to develop standards in the areas of dynamic spectrum access (DSA), cognitive radio (CR), interference management, coordination of wireless systems, and advanced spectrum management, among others [9].

Currently, there are seven Working Groups (WGs) within SCC 41. Of these, IEEE 1900.1[10], "Standard Definitions and Concepts for Dynamic Spectrum Access: Technology Relating to Emerging Wireless Networks, System Functionality, and Spectrum Management," was completed in September 2008, and IEEE 1900.2 [11], "Recommended Practice for the Analysis of In-Band and Adjacent-Band Interference and Coexistence Between Radio Systems," was approved by the IEEE Standards Board in July 2008. The 1900.3 WG has been disbanded. IEEE 1900.4 WG is for "Architectural Building Blocks Enabling Network-Device Distributed Decision Making for Optimized Radio Resource Usage in Heterogeneous Wireless Access Networks," and IEEE 1900.4 [12] was published on February 27th 2009. The P1900.5 WG is working on "Policy Language and Policy Architectures for Managing Cognitive Radio for Dynamic Spectrum Access Applications," and the standard [13] was published in January 2012. IEEE 1900.6 [14] "Spectrum Sensing Interfaces and Data Structures for Dynamic Spectrum Access and Other Advanced Radio Communication Systems," was published on April 22nd 2011. The recently added WG is 1900.7, "Radio Interface for White Space Dynamic Spectrum Access Radio Systems Supporting Fixed and Mobile Operation," and its draft standard is under development.

7.2.3.1   IEEE DySPAN-SC 1900.1

This standard provides definitions and explanations of key concepts in the fields of spectrum management, cognitive radio, policy-defined radio, adaptive radio, software-defined radio and related technologies. The document goes beyond simple, short definitions by providing text that explains these terms in the context of the technologies. The document also describes how these technologies interrelate and create new capabilities while at the same time providing mechanisms supportive of new spectrum management paradigms such as dynamic spectrum access.

From February 2nd 2011, the 1900.1 Working Group has been working on a new project: 1900.1a, It is an additive amendment of 1900.1, and it adds some new terms and associated definitions.

7.2.3.2   IEEE DySPAN-SC 1900.2

This recommended practice provides technical guidelines for analyzing the potential for coexistence or in contrast interference between radio systems operating in the same spectrum band or between different spectrum bands.

7.2.3.3   IEEE DySPAN-SC 1900.4

IEEE 1900.4 is intended to apply to a heterogeneous wireless environment. Such an environment may include multiple operators, multiple RANs, multiple RATs and multiple terminals. As the main subject of optimization, advanced spectrum management capabilities are considered in 1900.4. And an example of such a capability is that the assignment of spectrum to RANs can be dynamically changed, where "spectrum assignment" may be characterized as a carrier frequency, a signal bandwidth, or a radio interface to be used in the assigned spectrum. Another example of advanced spectrum management is that the assignment of spectrum to RANs is fixed, but some RANs are allowed to concurrently operate in more than one spectrum assignment. For backward compatibility, RANs in 1900.4 should be a legacy technology.

The underlying approach in 1900.4 is to define a management system that decides on a set of actions required to optimize the radio resource usage and quality of service in a heterogeneous wireless environment. In particular, 1900.4, in its first stage, defines the entities and interfaces of this management system. The two key management entities are shown in Fig. 7.2, including the network reconfiguration manager (NRM) and terminal reconfiguration manager (TRM). The composition of a heterogeneous wireless environment with 1900.4 management entities creates a composite wireless network, which includes various capabilities to optimize radio resource usage as well as the associated flexibility.

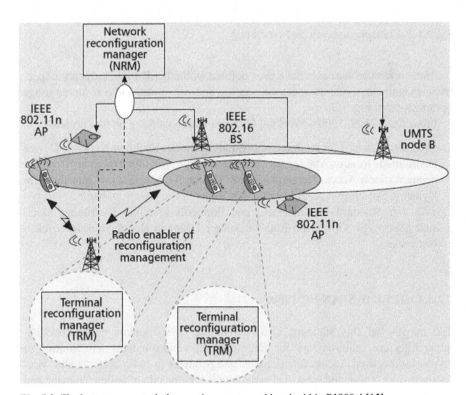

Fig. 7.2 The heterogeneous wireless environment considered within P1900.4 [15]

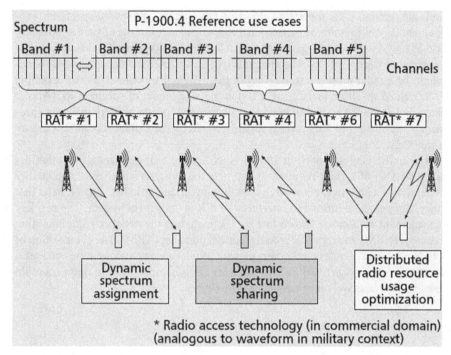

Fig. 7.3 The reference use cases for P1900.4 [15]

Three reference use cases have been defined within IEEE 1900.4: dynamic spectrum assignment, dynamic spectrum sharing and distributed radio resource usage optimization in Fig. 7.3.

From April 2009, 1900.4 Working Group has been working on two projects:

1. *1900.4A*: Standard for Architectural Building Blocks Enabling Network-Device Distributed Decision Making for Optimized Radio Resource Usage in Heterogeneous Wireless Access Networks—Amendment: Architecture and Interfaces for Dynamic Spectrum Access Networks in White Space Spectrum bands.
2. *1900.4.1*: Standard for Interfaces and Protocols Enabling Distributed Decision Making for Optimized Radio Resource Usage in Heterogeneous Wireless Networks.

7.2.3.4   IEEE DySPAN-SC 1900.5

The purpose of this standard is to define the policy language and associated architecture requirements for interoperable, vendor-independent control of Dynamic Spectrum Access functionality and behavior in radio systems and wireless networks. This standard will also define the relationship of policy language and architecture to the needs of at least the following constituencies: the regulator, the operator, the user, and the network equipment manufacturer.

The P1900.5 standard defines a vendor-independent set of policy-based control architectures and corresponding policy language requirements for managing the functionality and behavior of dynamic spectrum access networks.

P1900.5 standard was published in January 2012 [13]. Follow-on work for P1900.5.1 is in progress. There are three main projects:

1. *P1900.5.1*: Draft Standard Policy Language for Dynamic Spectrum Access Systems. Scope: This standard defines a vendor-independent policy language for managing the functionality and behavior of dynamic spectrum access networks based on the language requirements defined in the IEEE 1900.5 standard. The standard developed under PAR will take into consideration both the Policy Language Requirements of IEEE 1900.5 and the results of the Modeling Language for Mobility Work Group (MLM-WG) within the Wireless Innovation Forum (SDRF v2) Committee on Advanced Wireless Networking and Infrastructure. MLM-WG is developing use cases, an ontology, corresponding signaling plan, requirements and technical analysis of the information exchanges that enable next generation communications features such as spectrum awareness and dynamic spectrum adaptation, waveform optimization, capabilities, feature exchanges and advanced applications. The MLM-WG expects this effort to lead to specifications/standards for languages and data exchange structures to support these capabilities.
2. *P1900.5.a*: This will provide an amendment to P1900.5 defining the interface description between policy architecture components.
3. *P1900.5.2*: Standard Method for Modeling Spectrum Consumption. This standard defines a vendor-independent generalized method for modeling the RF spectrum consumption and the attendant computations for arbitrating the compatibility among models. The methods of modeling are chosen to support the development of tractable algorithms for determining the compatibility between models and for performing various spectrum management tasks that operate on a plurality of models. The modeling methods are exclusively focused on capturing spectrum use but are defined in a schema that can be joined with other schemata of business processes of spectrum management or behavioral aspects of spectrum policy.

7.2.3.5 IEEE DySPAN-SC 1900.6

This standard defines the interfaces and data structures required to exchange sensing-related information for increasing interoperability between sensors and their clients developed by different manufacturers. The clients can be cognitive engines as in the focus of this standard, or any other type of algorithms and devices (e.g., adaptive radio) that use sensing-related information. Being aware of evolving technologies, interfaces are developed to accommodate future extensions, new service

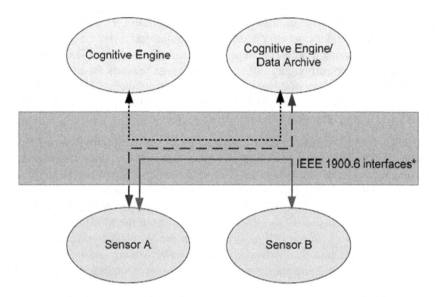

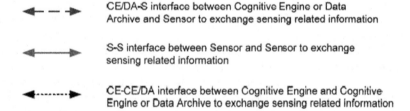

Fig. 7.4 IEEE 1900.6 interfaces between spectrum sensors and their clients [14]

primitives and parameters. This standard provides a formal definition of data structures and interfaces for the exchange of spectrum sensing-related information.

As shown in Fig. 7.4, there are three possible instances of the logical interface within the scope of this standard depending on the logical entities using the IEEE 1900.6 interface: (1) CE/DA–S2 interface between CE or data archive (DA) and sensor; (2) S–S interface between sensor and sensor; (3) CE–CE/DA interface between CE and CE or DA.

The CE/DA–S interface is used for exchanging sensing-related information between a CE or DA and a sensor. As an example, the CE/DA–S interface is used in scenarios where a given CE or DA obtains sensing-related information from one or several sensors or a given sensor provides sensing-related information to one or several CEs or DAs.

The S–S interface is used for exchanging sensing-related information between Sensors. As an example, the S–S interface is used in scenarios when multiple sensors exchange sensing-related information for distributed sensing.

The CE–CE/DA interface is used for exchanging sensing-related information between CEs or between CE and DA. The CE–CE/DA interface is used in scenarios where CEs exchange sensing-related information for distributed sensing. The CE–CE/DA interface is also used in scenarios where a CE obtains sensing-related information and or policy/regulatory information from a DA. For CE–CE/DA communication, a CE/DA shall be able to take the role of sensors in terms of providing the CE/DA–S interface for the duration and purpose of exchanging sensing-related information.

This amendment provides specifications to allow integrating 1900.6 based distributed sensing systems into existing and future dynamic spectrum access radio communication systems. It enables existing legacy systems to benefit so as to widen the potential adoption of the IEEE 1900.6 interface as an add-on to these systems and to claim standard conformance for an implementation of the interface. In addition, it facilitates sharing of spectrum sensing data and other relevant data among 1900.6 based entities and external data archives.

7.3 European Telecommunications Standards Institute

In January 2008, European Telecommunications Standards Institute (ETSI) Council # 65 approved the establishment of Technical Committee Reconfigurable Radio System (TC RRS). At the inaugural meeting, the ETSI RRS TC created the following four Working Groups (WGs) shown in Fig. 7.5, in which the technical discussions are organized and reports are produced [16].

Cognitive radio principles within ETSI RRS are concentrated on two topics, a cognitive pilot channel proposal [16] and a functional architecture [17] for management and control of reconfigurable radio systems, including dynamic self-organizing planning and management, dynamic spectrum management and joint radio resource management.

In order to define a functional architecture that is able to provide optimized management of radio and spectrum resources, WG3 of ETSI RRS has collected and reported the following set of requirements.

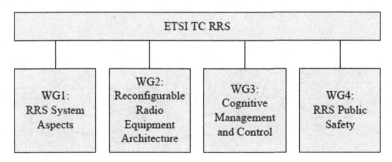

Fig. 7.5 ETSI RRS structure [16]

- Personalization, to support various classes of users
- Support of pervasive computing, enabled by the existence of sensors, actuators, and wireless local area networks in all application areas
- Context awareness, for efficiently handling multiple, dynamically changing, and unexpected situations
- Always best connectivity, for optimally serving equipment and users in terms of QoS and cost
- Ubiquitous application provision for the applications above
- Seamless mobility, for rendering the users agnostic of the heterogeneity of the underlying infrastructure
- Collaboration with alternate RATs for contributing to the achievement of always-best connectivity
- Scalability, for responding to frequent context changes

Accordingly, in order to address these requirements, a proper functional decomposition has been proposed in [17]. The derived functional blocks, together with the interfaces among them and their distribution between network and terminals, are depicted in Fig. 7.6. The dynamic spectrum management (DSM) block is responsible for the medium and long term, both technical and economical, management of spectrum, and as such it incorporates functionalities like provisioning of information for spectrum assignments, spectrum occupancy evaluation and decision making on spectrum sharing/trading.

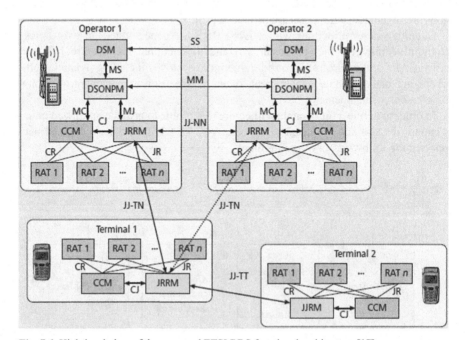

Fig. 7.6 High-level view of the proposed ETSI RRS functional architecture [17]

Dynamic self-organizing planning and management (DSONPM) caters for the medium and long-term management at the level of a reconfigurable network segment (e.g., incorporating several BSs). It provides decision making functionality for QoS assignments, traffic distribution, network performance optimization, RATs activation, configuration of radio parameters, and so on. The fundamental objective of the joint radio resources management (JRRM) block is the joint management of radio resources possibly belonging to heterogeneous RATs, and its functionalities mainly include radio access selection, neighborhood information provision, and QoS/bandwidth allocation/admission control. Moreover, configuration control module (CCM) is responsible for the enforcement of the reconfiguration decisions typically made by the DSONPM and JRRM.

7.4 Concluding Remarks

Currently, many standardizations related to CRS have proposed multiple deployment scenarios and business models. ITU-R 5A further develops the working document towards a report on cognitive radio systems in the land mobile service in accordance with Question ITUR 241-1/5 and Resolution ITU-R 58. Besides, IEEE 1900.7 is recently set up and defined the Radio Interface for white space dynamic spectrum access radio systems supporting fixed and mobile operation. This standard specifies a radio interface including medium access control (MAC) sublayer(s) and physical (PHY) layer(s) of white space for the dynamic spectrum access radio systems supporting fixed and mobile operation in white space spectrum bands, while avoiding harmful interference to incumbent users in these spectrum bands.

References

1. Resolution ITU-R 58 (2012) Studies on the implementation and use of cognitive radio systems
2. Recommendation 76 (WRC-12) (2012) Deployment and use of cognitive radio systems
3. Filin S, Murakami H, Harada H et al (2011) ITU-R standardization activities on cognitive radio systems. CROWNCOM, Osaka, 1–3 June 2011
4. Report ITU-R SM.2152 (2009) Definitions of software defined radio (SDR) and cognitive radio system (CRS)
5. Annex 5 to WP1B Chairman's Report 1B/267-E (2010) Draft CPM Text onWRC-12 Agenda Item 1.19
6. Report ITU-R M.2225 (2011) Introduction to cognitive radio systems in the land mobile service
7. Report ITU-R M.2242 (2011) Cognitive radio systems specific for international mobile telecommunications systems
8. Shin KG, Hyoil K, Min AW et al (2010) Cognitive radios for dynamic spectrum access: from concept to reality. IEEE Wireless Commun 17(6):64–74
9. IEEE DySPAN Standards Committee (2013) http://grouper.ieee.org/groups/dyspan/index.html. Accessed 21 Nov 2013

10. IEEE Standard (2008) Definitions and concepts for dynamic spectrum access: terminology relating to emerging wireless networks, system functionality, and spectrum management. IEEE Std 1900.1–2008

11. IEEE Standard (2008) Recommended practice for the analysis of in-band and adjacent band interference and coexistence between radio systems. IEEE Std 1900.2–2008

12. IEEE Standard (2009) Architectural building blocks enabling network-device distributed decision making for optimized radio resource usage in heterogeneous wireless access networks. IEEE Std 1900.4–2009

13. IEEE Standard (2012) Policy language requirements and system architectures for dynamic spectrum access systems. IEEE Std 1900.5–2011

14. IEEE Standard (2011) Spectrum sensing interfaces and data structures for dynamic spectrum access and other advanced radio communication systems. IEEE Std 1900.6–2011

15. Buljore S, Harada H, Filin S et al (2009) Architecture and enablers for optimized radio resource usage in heterogeneous wireless access networks: the IEEE 1900.4 working group. IEEE Commun Mag 47(1):122–129

16. ETSI TR 102 683 (2009) Reconfigurable radio systems (rrs)—cognitive pilot channel (CPC)

17. ETSI TR 102 682 (2009) Reconfigurable radio systems (RRS)—functional architecture for management and control of reconfigurable radio systems

Chapter 8
Conclusion and Future Research Directions

8.1 Concluding Remarks

The origin and challenge of CWN in terms of spectrum scarcity and heterogeneous network isolation problems are briefly introduced in this book. To characterize and quantize the information sequence of the radio environment awareness results, a novel concept of cognitive information theory has been proposed by using the geographic entropy and temporal entropy, which brings the intelligence to the traditional wireless network. Furthermore, the architecture of CWN is designed by integrating the cognition cycle functions to improve the efficiency of radio resource utilization and reduce the network overhead, which supports the heterogeneous network convergence. From the realistic implementation perspective, the efficient cognitive information delivery and storage technologies of cognitive pilot channel (CPC) and multi-domain cognition database are proposed, which can collect multi-domain information and enhance the cooperation among heterogeneous networks. In terms of the fluctuation feature of the available spectrum and diverse QoS requirements of various applications, the effective and intelligent dynamic spectrum management and joint radio resource management technologies are designed with numerous simulation results. To verify the proposed theories and technologies, the TD-LTE based CWN testbed has been designed and developed for the dynamic spectrum management and efficient spectrum utilization. As far as we know, it is the first prototype worldwide with advanced spectrum sensing and spectrum handover technologies with performance indicators and numerous results.

8.2 Potential Future Works

In face of the surge of user equipments and increasing demands of various service requirements, how to recognize the characteristics of user behavior and how to dynamically utilize the radio resource according to the demands in different time and geo-location domains are still unsolved yet. By adding the intelligent abilities to

Z. Feng et al., *Cognitive Wireless Networks*, SpringerBriefs in Electrical and Computer Engineering, DOI 10.1007/978-3-319-15768-9_8

wireless networks, the radio resource management can be improved extensively and adaptive to the changing service demands. Therefore, further researches are foreseen as depicted below.

1. Extraction of the characteristics of user behaviors in heterogeneous networks using data mining algorithms in terms of different time and location related effects.
2. The virtualization and software-defined functions in wireless access networks to improve the efficiency and scalability of network.
3. Intelligent wireless network environment awareness and effective cognitive information delivery technologies.
4. Theory and method to support dynamic topology reconstruction for wireless networks in terms of the dynamic changing number and location of users.

Printed in the United States
By Bookmasters